Quaternion Fourier Transforms for Signal and Image Processing

FOCUS SERIES

Series Editor Francis Castanié

Quaternion Fourier Transforms for Signal and Image Processing

Todd A. Ell
Nicolas Le Bihan
Stephen J. Sangwine

iSTE

WILEY

First published 2014 in Great Britain and the United States by ISTE Ltd and John Wiley & Sons, Inc.

ISTE Ltd
27-37 St George's Road
London SW19 4EU
UK

www.iste.co.uk

John Wiley & Sons, Inc.
111 River Street
Hoboken, NJ 07030
USA

www.wiley.com

Library of Congress Control Number: 2014934161

British Library Cataloguing-in-Publication Data
A CIP record for this book is available from the British Library
ISSN 2051-2481 (Print)
ISSN 2051-249X (Online)
ISBN 978-1-84821-478-1

MIX
Paper from responsible sources
FSC® C013604
FSC www.fsc.org

Printed and bound in Great Britain by CPI Group (UK) Ltd., Croydon, Surrey CR0 4YY

Contents

NOMENCLATURE ix

PREFACE xi

INTRODUCTION xiii

CHAPTER 1. QUATERNION ALGEBRA 1

1.1. Definitions 1
1.2. Properties 2
1.3. Exponential and logarithm of a quaternion 7
1.3.1. Exponential of a pure quaternion 7
1.3.2. Exponential of a full quaternion 9
1.3.3. Logarithm of a quaternion 10
1.4. Representations 11
1.4.1. Polar forms 11
1.4.2. The $\mathbb{C}_j$-pair notation 15
1.4.3. $\mathbb{R}$ and $\mathbb{C}$ matrix representations 17
1.5. Powers of a quaternion 18
1.6. Subfields 18

CHAPTER 2. GEOMETRIC APPLICATIONS 21

2.1. Euclidean geometry (3D and 4D) 21
2.1.1. 3D reflections 22
2.1.2. 3D rotations 22
2.1.3. 3D shears 24
2.1.4. 3D dilations 24

2.1.5. 4D reflections 25
2.1.6. 4D rotations 25
2.2. Spherical geometry 26
2.3. Projective space (3D) 28
2.3.1. Systems of linear quaternion functions 31
2.3.2. Projective transformations 33

CHAPTER 3. QUATERNION FOURIER TRANSFORMS 35

3.1. 1D quaternion Fourier transforms 38
3.1.1. Definitions 38
3.1.2. Basic transform pairs 40
3.1.3. Decompositions 42
3.1.4. Inter-relationships between definitions 45
3.1.5. Convolution and correlation theorems 47
3.2. 2D quaternion Fourier transforms 48
3.2.1. Definitions 48
3.2.2. Basic transform pairs 52
3.2.3. Decompositions 54
3.2.4. Inter-relationships between definitions 55
3.3. Computational aspects 57
3.3.1. Coding 57
3.3.2. Verification 62
3.3.3. Verification of transforms 62

CHAPTER 4. SIGNAL AND IMAGE PROCESSING 67

4.1. Generalized convolution 67
4.1.1. Classical grayscale image convolution filters 67
4.1.2. Color images as quaternion arrays 70
4.1.3. Quaternion convolution 70
4.1.4. Quaternion image spectrum 73
4.2. Generalized correlation 79
4.2.1. Classical correlation and phase correlation 81
4.2.2. Quaternion correlation 86
4.2.3. Quaternion phase correlation 88
4.3. Instantaneous phase and amplitude of complex signals 91
4.3.1. Important properties of 1D QFT of a complex signal $z(t)$ 91
4.3.2. Hilbert transform and right-sided quaternion spectrum 96
4.3.3. The quaternion signal associated with a complex signal 98
4.3.4. Instantaneous amplitude and phase 101
4.3.5. The instantaneous frequency of a complex signal 102

4.3.6. Examples 104
4.3.7. The quaternion Wigner-Ville distribution of a complex signal . . . 109
4.3.8. Time marginal 113
4.3.9. The mean frequency formula 113

BIBLIOGRAPHY 117

INDEX 123

Nomenclature

Symbol	Meaning	Section	Equation
$\mathbb{R}$	Set of real numbers	section 1.1	
$\mathbb{C}$	Set of complex numbers	section 1.1	
$\mathbb{H}$	Set of quaternions	section 1.1	
$\mathbf{V}(\mathbb{H})$	Set of pure quaternions	section 1.1	
$\mathbb{C}_{\boldsymbol{\mu}}$	Subfield of $\mathbb{H}$	section 1.6	
I	Complex root of -1	(See note below.)	
$\Re(.)$	Real part	section 1.1	
$\Im(.)$	Imaginary part	section 1.1	
$\boldsymbol{i}, \boldsymbol{j}, \boldsymbol{k}$	Quaternion basis elements	section 1.1	[1.2]
$\boldsymbol{\mu}$	Pure quaternion	section 1.1	[1.2]
$\Im_{\boldsymbol{i}}$	$\boldsymbol{i}$-imaginary part	section 1.1	[1.4]
$\Im_{\boldsymbol{j}}$	$\boldsymbol{j}$-imaginary part	section 1.1	[1.4]
$\Im_{\boldsymbol{k}}$	$\boldsymbol{k}$-imaginary part	section 1.1	[1.4]
$\mathcal{M}_{\mathbb{R}}(.)$	Real 4×4 matrix representation	section 1.4	[1.69]
$\mathcal{M}_{\mathbb{C}}(.)$	Complex 2×2 matrix representation	section 1.4	[1.70]
$[\![\cdot]\!]$	Alternate real 4×4 matrix representation	section 2.3	[2.20]
$[\cdot]$	Real 4×1 vector representation	section 2.3	[2.20]
$\mathrm{S}(.)$	Scalar part	section 1.1	[1.3]
$\mathbf{V}(.)$	Vector part	section 1.1	[1.3]
(q_1, q_2)	$\mathbb{C}_{\boldsymbol{j}}$-pair notation of q	section 1.4	
$\widehat{\mathcal{AB}}$	Arc of great circle between $\mathcal{A}$ and $\mathcal{B}$	section 2.2	
$\overline{q}$	Quaternion conjugate	section 1.2	[1.18]
$z^{\star}$	Complex conjugate	section 1.4	[1.54]
$\overline{q}^{\boldsymbol{\mu}}$	Involution	section 1.2	[1.23]
q^{-1}	Inverse	section 1.2	[1.26]
$\langle p, q \rangle$	Inner product	section 1.2	[1.10]
$\boldsymbol{\mu} \times \boldsymbol{\eta}$	Cross product	section 1.2	[1.9]
$\boldsymbol{\mu} \perp \boldsymbol{\eta}$	Orthogonality	section 1.2	

$q \odot p$	Bicomplex product	section 1.4	[1.65]
$\{1, \boldsymbol{i}, \boldsymbol{j}, \boldsymbol{k}\}$	Canonical basis in $\mathbb{H}$	section 1.1	
$\|q\|$	Norm	section 1.2	[1.14]
$|q|$	Modulus	section 1.2	[1.16]
$\tilde{q}$	Unit quaternion $|\tilde{q}| = 1$	section 1.2	[1.30]
$f * g$	Convolution	section 3	[3.1]
$f \star g$	Correlation	section 3	[3.1]
$\mathcal{F}\{.\}$	Fourier transform	section 3	[3.1]
$\mathcal{P}_{\boldsymbol{\eta}}[\cdot]$	Reflection operator	section 2.1	[2.1]
$\mathcal{R}_q[\cdot]$	Rotation operator	section 2.1	[2.3]
$\mathcal{S}_{\alpha,\beta,\boldsymbol{\mu}}[\cdot]$	Shear operator	section 2.1	[2.9]
$\mathcal{D}_{\boldsymbol{\mu},\alpha}[\cdot]$	Dilation operator	section 2.1	[2.10]
$L^1(\mathbb{G};\mathbb{K})$	Space of absolutely integrable $\mathbb{K}$-valued functions taking arguments in $\mathbb{G}$	section 3.1	
$L^2(\mathbb{G};\mathbb{K})$	Space of square integrable $\mathbb{K}$-valued functions taking arguments in $\mathbb{G}$	section 4.3	
$\mathrm{sgn}(.)$	Signum function	section 4.3	[4.31]
$p.v.(.)$	Principal value of an integral	section 4.3	[4.32]

NOTE.– The complex root of -1 which is usually denoted i, or, in engineering texts j, is denoted throughout this book by a capital letter I, in order to avoid any confusion with the first of the three quaternion roots of -1, all three of which are denoted throughout in bold font like this: $\boldsymbol{i}$, $\boldsymbol{j}$, $\boldsymbol{k}$.

Preface

This book aims to present the state of the art, together with the most recent research results in the use of quaternion Fourier transforms (QFTs) for the processing of color images and complex-valued signals. It is based on the work of the authors in this area since the 1990s and presents the mathematical concepts, computational issues and some applications to signals and images. The book, together with the MATLAB® toolbox developed by the authors, [SAN 13b] allows the readers to make use of the presented concepts and experiment with them in practice through the examples provided.

Todd A. ELL
Minneapolis, MN, USA

Nicolas le BIHAN
Melbourne, Australia

Stephen J. SANGWINE
Colchester, UK

April 2014

Introduction

This book covers a topic that combines two branches of mathematical theory to provide practical tools for the analysis and processing of signals (or images) with three- or four-dimensional samples (or pixels). The two branches of mathematics are not recent developments, but their combination has occurred only within the last 25–30 years, and mostly since just before the millennium.

I.1. Fourier analysis

Fourier analysis was, in 1822, with Joseph Fourier's development of techniques, the first to analyze mathematical functions into sinusoidal components. In signal and image processing, Fourier's ideas underpin the two fundamental representations of a signal: one in the *time (or image) domain* where the signal (or image) is represented by samples (or pixels) with amplitudes and the other in the *frequency domain* where the signal (or image) is represented by sinusoidal frequency components, each with an amplitude and a phase. Mathematically, these concepts are not limited to time and frequency: one can use Fourier analysis on a function of any variable, resulting in a representation in terms of sinusoidal functions of that variable. However, this book is concerned with signal and image processing, and we will therefore use the terms *time* and *frequency* rather than more general concepts. It should be understood throughout that when we talk of images, the concept of time is replaced by the two spatial coordinates that define pixel position within an image.

Today, Fourier analysis is classically taught to mathematicians, scientists and engineers in several related ways, each applicable to a specific subset of mathematical functions or signals:

1) Fourier *series* analysis [SNE 61] in which continuous *periodic* functions of time, with infinite duration, are represented as sums of cosine and sine functions, each with infinite duration;

– Fourier *integrals* or transforms [BRA 00, ROB 68] in which continuous (but aperiodic) functions of time are represented as continuous functions of frequency (or *vice versa*);

2) Discrete Fourier transforms in which signals defined at discrete intervals in time are represented in the frequency domain by cosine and sine functions. This topic is broken down into:

– discrete-time Fourier transforms, in which discrete-time signals of limited duration are represented as continuous frequency-domain distributions;

– discrete Fourier transforms, in which discrete-time, discretized (that is *digital*) signals of finite duration are represented by a finite-length array of digital frequency coefficients. (These are usually computed numerically using the *fast Fourier transform* (FFT)).

The key to all of the above ideas is the representation of a signal using complex exponentials, often known as *harmonic analysis*, although this term has a somewhat wider meaning in mathematics than its usage in signal and image processing. The complex exponential with angular frequency ω and phase ϕ: $f(t) = A\exp(\omega t + \phi) = A\left(\cos(\omega t + \phi) + I\sin(\omega t + \phi)\right)$ has cosine and sine components in its real and imaginary parts, respectively. Since, in this book, we are concerned with signals that have three- or four-dimensional samples, it is helpful to consider classical Fourier analysis in terms of complex exponentials rather than in terms of separate cosines and sines.

Figure I.1 shows a real-valued signal (on the left-hand side of the plot, with time increasing away from the viewer). The signal is a sawtooth waveform reconstructed from its first five non-zero harmonics, which are plotted in the center of the figure as helices. (The horizontal spacing between the helices is introduced simply to make them clearer: there is no mathematical significance to it). The five helices on the left are the positive frequency complex exponentials and the five helices on the right are the negative frequencies. Note that the positive and negative frequency exponentials have opposite directions of rotation. The real parts of the harmonics are projected onto the right-hand side of the figure (these sum to give the reconstructed waveform on the left) and the imaginary parts of the harmonics are projected onto the base of the figure (these cancel out because the exponentials occur in complex conjugate pairs at positive and negative frequencies, a symmetry due to the original signal being real-valued).

In general, with a complex signal analyzed into complex exponentials in the same way, there would be no symmetry between the positive and negative frequency exponentials. This case is a useful model for what follows in this book, where we consider signals and images with three- and four-dimensional samples. Figure I.2 shows a complex signal constructed by bandlimiting a random complex signal.

Time is plotted on the right, increasing to the right, and at each time instant the signal has a complex value. The signal evolution over time traces out a path in the complex plane, and the figure renders this path as a three-dimensional view by plotting the signal values, in effect, on a stack of 2,000 transparent complex planes perpendicular to the time axis. The real and imaginary parts of the signal are also plotted on the base of the axes, and on the rear plane of the axes. Analysis of a complex signal into positive and negative frequency complex exponentials is not conceptually different from the real case depicted in Figure I.1: each complex exponential will have an amplitude and phase, and their sum will reconstruct the original signal.

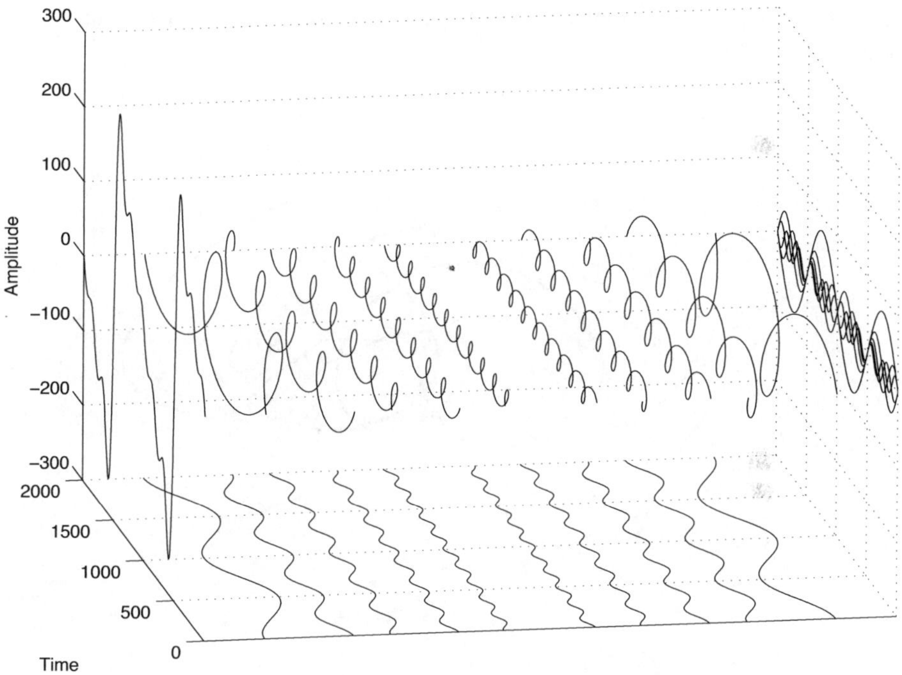

Figure I.1. *Analysis of a real signal into complex exponential harmonics*

The time and frequency domain representations of a signal are not mutually exclusive: the field of time-frequency analysis [FLA 98] is concerned with intermediate representations that combine aspects of time and frequency. The need for intermediate representations arises due to the variation of frequency content in a signal over time. This is not an easy concept to understand, but it follows from the uncertainty principle or *Gabor limit*: a signal cannot be bandlimited (i.e. with frequency content limited to a finite range of frequencies) and simultaneously be of limited time duration. A pure sinusoidal signal with unlimited duration (infinite extent) can be represented in the frequency domain as an impulse (that is a function with zero value everywhere except at one frequency point). Conversely, an impulse in the time domain has infinite bandwidth. However, a signal that contains a specific frequency for a limited time requires a time-frequency representation. Examples of such signals occur widely in the real world: speech and music contain frequencies that are present for a short time (one note played on a musical instrument, for example, which lasts for the duration of the note, plus some reverberation time afterward). An in-depth discussion of these ideas is outside the scope of this book, but is assumed to be understood; although much of the contents of the book relates to Fourier transforms, the quaternion approach can easily be applied to time-frequency concepts, such as fractional and short-time Fourier transforms, by combining quaternion transform formulations with existing knowledge from classical signal processing.

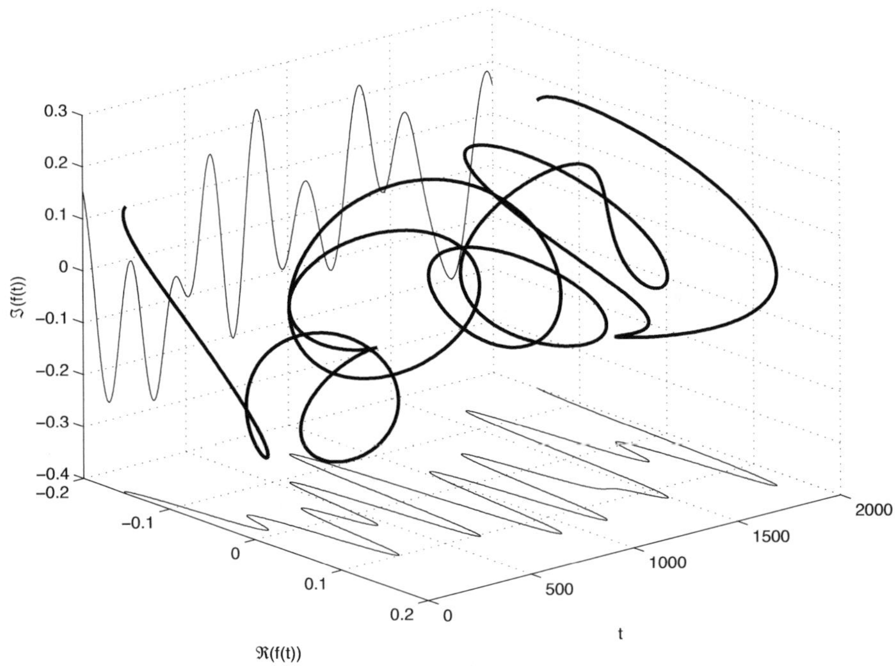

Figure I.2. *A bandlimited complex signal showing real and imaginary parts projected onto the base and rear of the grid box*

I.2. Quaternions

In this book, we are concerned with signals and images that have vector-valued samples (that is samples with three or more components), and their processing using

Fourier transforms based on four-dimensional hypercomplex numbers (quaternions). In Chapter 4 (section 4.3), we show that quaternion Fourier transforms also have applications for the processing of complex signals, exploiting the symmetry properties of a quaternion Fourier transform that are missing from a complex Fourier transform.

A vector-valued signal (in three dimensions, for example) evolves over time and traces out a path in three-dimensional space. To render a plot of such a signal requires four dimensions, and therefore we cannot produce a graphical representation like the one in Figure I.2. Decomposition of a vector-valued signal into harmonic components requires a Fourier transform in an algebra with dimension higher than 2, and this is the motivation for the use of quaternions, which, as we will see, are the next available higher-dimensional algebra after the complex numbers.

Quaternions followed the work of Fourier just over 20 years later, Sir William Rowan Hamilton in 1843 to generalize the complex numbers to three dimensions, was forced to resort to four dimensions in order to obtain what we now call a *normed division algebra*, that is, an algebra where the norm of a product equals the product of the norms, and where every element of the algebra (except zero) has a multiplicative inverse [WAR 97]. Hamilton opened a door in mathematics to hypercomplex algebras in general [STI 10, Chapter 20], [KAN 89], leading to the octonions [CON 03, BAE 02] in less than a year, and the Clifford algebras about 30 years later [LOU 01, POR 95].

I.3. Quaternion Fourier transforms

Quaternion Fourier transforms, the subject of this book, are a generalization of the classical Fourier transform to process signals or images with three- or four-dimensional samples. Such signals arise very naturally in the physical world from the three dimensions of physical space. Quite independently, for very different (physiological) reasons connected with the trichromatic nature of human color vision [MCI 98], color images have three components per pixel. The fourth dimension of the quaternions plays a role in at least two ways: the frequency-domain representation of a signal with three-dimensional samples requires four dimensions (just as in the complex case, two dimensions are required in the frequency domain, even if the original signal has one-dimensional samples). But more generally, the four dimensions of the quaternions can be used to represent a most general set of geometric operations in three dimensions using homogeneous coordinates, which are explored in a later chapter (see section 2.3) and in [SAN 13a]. Of course, generalizations to higher dimensions are possible, and there is a wide range of work on Clifford Fourier transforms, which is outside the scope of this book (we refer the readers to a recent volume for further details [HIT 13], and in particular the historical introduction contained within [BRA 13]).

I.4. Signal and image processing

Fourier transforms are a fundamental tool in signal and image processing. They convert a signal or image from a representation based on sample or pixel amplitudes into a representation based on the amplitudes and phases of sinusoids. The latter representation is said to be in the *frequency domain*, and the original signal is said to be in the *time domain* for a signal which is a time series, or in the *image domain* for an image captured with a camera or scanner. Of course, signals may be encountered that are not time series, for example, measurements of some physical quantity made at (regular) intervals in space; in this book, we will use the terminology of time series for simplicity, since the processing of other signals is mathematically no different.

The Fourier transformation is *invertible*, which means that the original signal or image may be recovered from the frequency domain representation. More interestingly, the frequency domain representation may be modified before inversion of the transform, so that the recovered signal or image is a modified version of the original, for example, with some frequencies or bands of frequencies suppressed, attenuated or amplified. In some applications, inversion of the transform is not needed: the processing performed in the frequency domain directly yields information that can be immediately utilized. An example is computer vision, where a decision based on analysis of an image may result in an action without any need to construct an image from the processed frequency domain representation. At a more detailed level, another example includes correlation, where the signal or image is processed in the frequency domain to yield information about the location of a known object within an image (the same applies in signal processing to find a known signal occurring within a longer, noisy signal).

The classical Fourier transform is inherently based on complex numbers. This is obvious from the fact that the frequency domain representation must represent both the amplitude and the phase of each frequency present in the signal or image. The symmetry of the transform means that the signal may be complex without any modification of the transform. (There are some specialized variants of the Fourier transform that handle only real signals, for example the Hartley transform [BRA 86]). Given a signal with three components (representing, for example, acceleration in three mutually perpendicular directions), how can a frequency domain representation be calculated? The question is very similar if one considers a color image: is it possible to construct a *holistic* frequency-domain representation of the entire image? Obviously one can compute separate classical (i.e. complex) Fourier transforms of the three components in both of these cases, but one then has three separate frequency-domain representations, each representing one aspect of the original image (the frequency content of one of the color or luminance/chrominance channels). Processing of separate representations is sometimes known as *marginal* processing, for reasons connected with techniques in the hand computation of *marginal distributions* in statistics [TRU 53, section 1.22]. It is axiomatic in this book that marginal processing is not the best way to handle signals and images with more than two components per sample, but we will attempt to justify this belief throughout the book, by showing how holistic approaches with quaternions yield better results.

There is another reason for using a quaternion Fourier transform in some applications, and it provided the motivation for the earliest published work on quaternion Fourier transforms (in the field of nuclear magnetic resonance (NMR)). When a two-dimensional signal is captured (that is samples are measured over a two-dimensional grid, like an image), it is sometimes necessary to regard the two dimensions of the sampling grid as independent time-like axes. Processing such a signal with a classical two-dimensional complex transform mixes the two dimensions, whereas a suitably formulated quaternion transform does not. This is because it is possible to associate each of the time-like dimensions with a different dimension of the four-dimensional quaternion space, thus keeping the frequency-domain representations of the first and second time-like axes apart. There were two independent (as far as we are aware) early formulations of quaternion Fourier transforms, by Ernst [ERN 87, section 6.4.2] and Delsuc [DEL 88, equation 20], which are almost equivalent (they differ in the relative placement of the exponentials and the signal, and in the signs, inside the exponentials):

$$F(\omega_1, \omega_2) = \int_{-\infty}^{\infty} \int_{-\infty}^{\infty} f(t_1, t_2) e^{\boldsymbol{i}\omega_1 t_1} e^{\boldsymbol{j}\omega_2 t_2} \mathrm{d}t_1 \mathrm{d}t_2 \qquad \text{[I.1]}$$

Note that the two-dimensional signal $f(t_1, t_2)$, is scalar-valued (i.e. not quaternion-valued as in most of the cases discussed in this book). The two time-like axes t_1 and t_2 are treated using separate quaternion roots of -1 ($\boldsymbol{i}$ and $\boldsymbol{j}$), and therefore they are not mixed. This may also be regarded as being due to the orthogonality of the imaginary parts of the two exponentials.

Similar considerations motivated Thomas Bülow [BÜL 99, BÜL 01] when he processed grayscale images using a quaternion Fourier transform. By using a transform with samples of dimension greater than 2, Bülow was able to study symmetries present in certain images in a way that is not possible with the two-dimensional complex Fourier transform.

I.5. Other hypercomplex algebras

Of course, there are alternatives to quaternions for the construction and computation of Fourier transforms for the type of applications suggested so far in this introductory chapter, and it is worth briefly reviewing them here in order to provide a full picture. As already noted, the quaternion algebra is not the only hypercomplex algebra. In fact, the quaternion algebra is a specific example of a more general class of hypercomplex algebras discovered by William Kingdon Clifford in 1876 (about 30 years after Hamilton's discovery of quaternions) and known as *Clifford algebras*. The Clifford algebras include the complex numbers and quaternions, but not the octonions, curiously. However, as already mentioned, the quaternions share with the real numbers and the complex numbers a very specific property that sets them aside from all other Clifford algebras: every non-zero quaternion, $q \neq 0$, has a multiplicative inverse such that $q\,q^{-1} = q^{-1}q = 1$. Furthermore, the quaternion algebra is *normed*. This means that it is possible to define a norm (representing the squared length of the quaternion in four dimensions) such that the norm of a product of two quaternions equals the product of the norms of the two quaternions taken separately: $\|pq\| = \|p\|\,\|q\|$. This is discussed in section 1.2 (see [1.14]). Hypercomplex algebras in general have other troublesome properties. Many contain values that are *idempotent* or *nilpotent*. An idempotent value q squares to give itself: $q^2 = q$; and a nilpotent value squares to give zero: $q^2 = 0$. Such values obviously have the potential to cause problems in numerical algorithms [ALF 07], and the choice of the quaternion algebra avoids these problems entirely, because there are no nilpotent or idempotent quaternions other than 0 and 1, respectively.

The one property of the quaternions that cannot be avoided is that multiplication of quaternions is not commutative. This means that pq gives a different result in general from qp for two arbitrary quaternions p and q. The reason for this can be stated quite simply – the vector (or cross) product in three dimensions is not commutative, and the product of two quaternions contains a vector product. It is important to understand that this is an inherent property of three-dimensional geometry and is not specific to the quaternions. This is again discussed in Chapter 1. Non-commutative multiplication can be avoided by choosing a different hypercomplex algebra, but since any hypercomplex algebra which is commutative contains divisors of zero (a consequence of the Frobenius [DIC 14, section 11]), any attempt to avoid non-commutative multiplication will inevitably lead to other problems, which may well be more troublesome than non-commutativity. Non-commutative multiplication also occurs in linear algebra, of course, where the product of matrices is dependent on ordering; so it should not cause undue concern to anyone contemplating using quaternions.

I.6. Practical application

The ideas and concepts in this book are realisable in practice in several different ways, particularly using software.

I.6.1. *Software libraries*

The library [SAN 13b] permits experimentation with transforms and other algorithms operating on three- and four-dimensional data in MATLAB®. Since

MATLAB® can generate C and C++ code (with some restrictions on supported language features), quaternion code can, in principle, be used to generate code for stand-alone applications, subject to licensing[1].

Alternatively, code can be custom-written, using a quaternion library such as the Boost library for C++, which contains some quaternion functions in the Math toolkit [BRI 13]. This is discussed again in section 3.3, particularly with respect to the use of decompositions into complex transforms to avoid the need to code elaborate algorithms directly in quaternion code.

Both the QFTM and Boost libraries adopt the approach of directly coding quaternion operations, that is they represent quaternions as quadruplets of real (or complex) values, and provide elementary functions to add and multiply quaternions, implementing the famous rules for $\boldsymbol{ijk}$ given in section 1.1, directly in code.

I.6.2. *Matrix representations*

An alternative to the use of quaternion libraries is possible, using matrix representations, which we discuss here, and will return to with a very practical application (to verification) in section 3.3.3.1.

Hypercomplex algebras (with the exception of the octonions [CON 03, BAE 02], which are not associative) have matrix representations. What this means is that for a given algebra, there exists a matrix algebra with real or complex elements that is equivalent to the given hypercomplex algebra, in the sense that multiplication (and addition of course) of the matrix representations is equivalent to multiplication in the hypercomplex algebra. There are also other equivalences, for example the norm of a hypercomplex value may be equivalent to the determinant of the matrix representation. The matrix algebra using the given representation is said to be *isomorphic* to the hypercomplex algebra. Matrix representations for the quaternions are discussed in section 1.4.3, but we discuss the ramifications here.

Given the existence of a matrix representation of quaternions, it is theoretically possible to substitute matrix representations for quaternions, both in algebraic manipulation and in computer coding (the same is true for other hypercomplex algebras except the octonions). Doing so can be a useful technique in theoretical development because it can reduce a hypercomplex problem to a problem involving real (or complex) matrices, and thus provide a deep insight into the relation between the hypercomplex case and the well understood real and complex cases. However, there are disadvantages of using a matrix representation compared to a direct quaternion approach, *in practice*:

1 QTFM is licensed under the GNU General Public License.

1) Computation with matrices is numerically inefficient, and it requires four times as much memory as a direct quaternion representation storing only four values. A quaternion product requires 16 multiplications and 12 additions, whereas the equivalent 4×4 matrix product requires 64 multiplications and 48 additions, an increase by a factor of 4. This disadvantage effectively rules out the use of matrix representations for implementation, even for computer simulations (as run-time is four times less using a quaternion library that directly calculates quaternion products than using a general matrix package and a matrix representation for each quaternion).

2) The result of a sequence of arithmetic operations *in matrix form* may not be an accurate representation of a quaternion matrix. This disadvantage again rules out the use of matrix representations for implementation (a quaternion library will yield more accurate results).

3) The matrix representation provides little geometric insight. As will be shown in Chapter 2, the quaternion algebra provides a remarkably intuitive link with the geometry of three or four dimensions. It is possible to manipulate quaternion symbols algebraically in order to derive expressions for geometric operations. This is the central idea in *geometric algebras* [SOM 01]. Geometric algebras, as might be expected from the preceding text, are not division algebras; so there is a price to be paid for their additional geometric utility in the form of divisors of zero, which make them less attractive for applications in digital signal and image processing.

The matrix representation is certainly useful, and it is helpful in any quaternion library to have the ability to convert between a direct quaternion representation and the matrix representation. In the QTFM library [SAN 13b], for example, functions called *adjoint*[2] and *unadjoint* are provided to perform the conversion, even for matrices of quaternions (the adjoint in the latter case is a block matrix with each block representing one quaternion).

I.7. Overview of the remaining chapters

The rest of the book is divided into four chapters. Chapter 1 covers the quaternion algebra, and provides the mathematical definitions and concepts necessary for the later chapters. Chapter 2 presents the geometric applications of quaternions, and provides the ideas necessary to understand how quaternions can be used to represent both three- and four-dimensional values and geometric operations applied to them. Chapter 3 gives a detailed and comprehensive account of quaternion Fourier transforms, including their definitions, operator formulas and how they may be computed. Chapter 4 shows how quaternion Fourier transforms can be applied in signal and image processing.

2 The "adjoint" terminology was taken from Zhang's 1997 paper [ZHA 97].

1

Quaternion Algebra

This chapter introduces the quaternion algebra $\mathbb{H}$ and presents some properties that will be useful in later chapters.

1.1. Definitions

Quaternions are one of the four existing normed division algebras over the real numbers. Classically denoted by $\mathbb{H}$ in honor of Sir W.R. Hamilton who discovered them in 1843, they form a non-commutative algebra. A quaternion $q \in \mathbb{H}$ is a four-dimensional (4D) hypercomplex number and has a Cartesian form given by:

$$q = a + \boldsymbol{i}b + \boldsymbol{j}c + \boldsymbol{k}d \qquad [1.1]$$

where $a, b, c, d \in \mathbb{R}$ are called its *components*. The three imaginary units $\boldsymbol{i}, \boldsymbol{j}, \boldsymbol{k}$ are square roots of -1 and are related through the famous[1] relations:

$$\begin{aligned} \boldsymbol{i}^2 &= \boldsymbol{j}^2 = \boldsymbol{k}^2 = \boldsymbol{ijk} = -1 \\ \boldsymbol{ij} &= -\boldsymbol{ji} = \boldsymbol{k} \\ \boldsymbol{ki} &= -\boldsymbol{ik} = \boldsymbol{j} \\ \boldsymbol{jk} &= -\boldsymbol{kj} = \boldsymbol{i} \end{aligned} \qquad [1.2]$$

A quaternion $q \in \mathbb{H}$ can be decomposed into a scalar part $\mathrm{S}(q)$ and a vector part $\mathbf{V}(q)$:

$$q = \mathrm{S}(q) + \mathbf{V}(q) \qquad [1.3]$$

1 These relations defining the three imaginary units of an element of $\mathbb{H}$ were carved by Sir W. R. Hamilton on a stone of the Broome bridge in Dublin on 16 October 1843.

where $\mathrm{S}(q) = a$ and $\mathbf{V}(q) = q - \mathrm{S}(q) = \boldsymbol{i}b + \boldsymbol{j}c + \boldsymbol{k}d$. Obviously, $\mathrm{S}(q) \in \mathbb{R}$ and we will also refer to it as the real part of q, i.e. $\Re(q) = a$. Now, $q \in \mathbb{H}$ will be called a *pure quaternion* if its real part is null, i.e. if $\mathrm{S}(q) = 0$. The set of pure quaternions will be denoted as $\mathbf{V}(\mathbb{H})$, while the set of quaternions with a null vector part are trivially identified with elements of $\mathbb{R}$, i.e. $\mathrm{S}(\mathbb{H}) \equiv \mathbb{R}$. If q has a null vector part, $\mathbf{V}(q) = 0$, then q is simply an element of $\mathbb{R}$. To identify the different imaginary components of a quaternion $q = a + \boldsymbol{i}b + \boldsymbol{j}c + \boldsymbol{k}d$, we will use the following notations:

$$\begin{cases} \Im_{\boldsymbol{i}}(q) = b \\ \Im_{\boldsymbol{j}}(q) = c \\ \Im_{\boldsymbol{k}}(q) = d \end{cases} \qquad [1.4]$$

so that a quaternion $q \in \mathbb{H}$ can be written as:

$$q = \Re(q) + \boldsymbol{i}\Im_{\boldsymbol{i}}(q) + \boldsymbol{j}\Im_{\boldsymbol{j}}(q) + \boldsymbol{k}\Im_{\boldsymbol{k}}(q) \qquad [1.5]$$

The Cartesian notation for a quaternion $q \in \mathbb{H}$ is, in fact, its expression in a specific 4D basis of the algebra $\mathbb{H}$, namely in the basis $\{1, \boldsymbol{i}, \boldsymbol{j}, \boldsymbol{k}\}$. Recall that, as an algebra, $\mathbb{H}$ possesses a vector space structure that allows the expression of any of its elements in terms of its *components* in a basis of $\mathbb{H}$. The basis $\{1, \boldsymbol{i}, \boldsymbol{j}, \boldsymbol{k}\}$ is the most common and popular basis to express a quaternion. However, we may encounter some other bases later on in this book, leading to alternate notations for a quaternion $q \in \mathbb{H}$. Before introducing these notations, we first review some remarkable properties of quaternions.

1.2. Properties

Here, we list some of the properties of quaternions that will be used throughout the book.

From the algebra structure of $\mathbb{H}$, the sum of two quaternions is trivial. Given two quaternions q and p, we have:

$$q + p = \mathrm{S}(q) + \mathrm{S}(q) + \mathbf{V}(p) + \mathbf{V}(q) \qquad [1.6]$$

Expressing the two quaternions in their Cartesian forms, $q = a + \boldsymbol{i}b + \boldsymbol{j}c + \boldsymbol{k}d$ and $p = e + \boldsymbol{i}f + \boldsymbol{j}g + \boldsymbol{k}h$, their sum is:

$$q + p = (a + e) + \boldsymbol{i}(b + f) + \boldsymbol{j}(c + g) + \boldsymbol{k}(d + h) \qquad [1.7]$$

and their product takes the form:

$$\begin{aligned} qp &= (a + \boldsymbol{i}b + \boldsymbol{j}c + \boldsymbol{k}d)\,(e + \boldsymbol{i}f + \boldsymbol{j}g + \boldsymbol{k}h) \\ &= ae - (bf + cg + dh) + \\ &\quad a\,(\boldsymbol{i}f + \boldsymbol{j}g + \boldsymbol{k}h) + e\,(\boldsymbol{i}b + \boldsymbol{j}c + \boldsymbol{k}d) + \\ &\quad \boldsymbol{i}\,(ch - dg) + \boldsymbol{j}\,(df - bh) + \boldsymbol{k}\,(bg - cf) \end{aligned} \qquad [1.8]$$

Using the scalar/vector notation, this product takes the following form:

$$qp = \mathrm{S}(q)\,\mathrm{S}(p) - \langle \mathbf{V}(q)\,, \mathbf{V}(p)\rangle + \mathrm{S}(q)\,\mathbf{V}(p) + \mathrm{S}(p)\,\mathbf{V}(q) + \mathbf{V}(q) \times \mathbf{V}(p) \qquad [1.9]$$

where $\langle .,. \rangle$ is the scalar product and $\times$ is the vector cross product. These are understood in the classical sense of the three-dimensional (3D) vector cross and inner products, which means that:

$$\langle \mathbf{V}(q)\,, \mathbf{V}(p)\rangle = bf + cg + dh \qquad [1.10]$$

which is scalar valued, i.e. $\langle \mathbf{V}(q)\,, \mathbf{V}(p)\rangle \in \mathbb{R}$, and that:

$$\mathbf{V}(q) \times \mathbf{V}(p) = \boldsymbol{i}(ch - dg) + \boldsymbol{j}(df - bh) + \boldsymbol{k}(bg - cf) \qquad [1.11]$$

where the result is a pure quaternion, i.e. $(\mathbf{V}(q) \times \mathbf{V}(p)) \in \mathbf{V}(\mathbb{H})$.

A very noticeable property is that the product of two quaternions is not commutative so that in general:

$$qp \neq pq \qquad [1.12]$$

This can be inferred from the presence of the non-commutative cross product in [1.9]. Note, however, that the product of quaternions is associative so that for any three quaternions $q, p, r \in \mathbb{H}$, the following is true:

$$(qp)\,r = q\,(pr) \qquad [1.13]$$

The norm of a quaternion q is defined as:

$$\|q\| = a^2 + b^2 + c^2 + d^2 \qquad [1.14]$$

A quaternion $q \in \mathbb{H}$ with $\|q\| = 1$ is said to be a *unit quaternion*. As previously mentioned, $\mathbb{H}$ is one of the four existing normed division algebras. As a result, given any two quaternions $p, q \in H$, then:

$$\|qp\| = \|q\|\,\|p\| \qquad [1.15]$$

It can also be easily checked that $\|qp\| = \|pq\|$. A related quantity that will be used in the following is the *modulus* of a quaternion. It is defined as the length of the quaternion in Euclidean 4D space. The modulus of $q \in \mathbb{H}$ is denoted by $|q|$ and is expressed as:

$$|q| = \sqrt{a^2 + b^2 + c^2 + d^2} = \|q\|^{\frac{1}{2}} \quad [1.16]$$

Obviously, $|q| \in \mathbb{R}^+$ and $|q| = 0$ if and only if $q = 0$. Like the norm, the modulus of a product of two quaternions p and q has the following property:

$$|pq| = |p|\,|q| = |qp| \quad [1.17]$$

Just as with the complex numbers, the conjugate of a quaternion q is obtained by negating its imaginary part. However, in $\mathbb{H}$ the imaginary part is 3D and consists of the entire vector part $\mathbf{V}(q)$. Denoted by $\overline{q}$, the conjugate of q is thus defined as:

$$\overline{q} = a - \boldsymbol{i}b - \boldsymbol{j}c - \boldsymbol{k}d = \mathrm{S}(q) - \mathbf{V}(q) \quad [1.18]$$

It follows that the scalar and vector parts of any quaternion $q \in \mathbb{H}$ can be obtained by:

$$\mathrm{S}(q) = \tfrac{1}{2}\,(q + \overline{q}) \quad \text{and} \quad \mathbf{V}(q) = \tfrac{1}{2}\,(q - \overline{q}) \quad [1.19]$$

Conjugation in $\mathbb{H}$ has the following property:

$$\overline{\overline{q}} = q \quad [1.20]$$

In contrast to the complex case, conjugation is not an involution[2] but an anti-involution, such that for $q, p \in \mathbb{H}$:

$$\overline{qp} = \overline{p}\,\overline{q} \quad [1.21]$$

that is, the order of the factors in a quaternion product is reversed by the conjugation operator. Note that the modulus (and also the norm) of a quaternion q can be expressed using the conjugate of q as:

$$|q| = \sqrt{q\overline{q}} \quad \text{and} \quad \|q\| = q\overline{q} \quad [1.22]$$

2 An involution $f : \mathbb{H} \to \mathbb{H}$ is such that for $q, p \in \mathbb{H}$:

$$f(f(q)) = q$$
$$f(p + q) = f(p) + f(q)$$
$$f(pq) = f(p)f(q)$$

It is also possible to define involutions over $\mathbb{H}$. Involutions are defined with respect to a pure unit quaternion $\boldsymbol{\mu}$ ($\boldsymbol{\mu} \in \mathbf{V}(\mathbb{H})$ and $|\boldsymbol{\mu}| = 1$). The most general case (together with many properties) is presented in [ELL 07d]. As special cases, one can choose $\boldsymbol{\mu}$ as one of the standard basis elements of $\mathbb{H}$, i.e. $\boldsymbol{i}$, $\boldsymbol{j}$ or $\boldsymbol{k}$. Involutions with respect to these three unit pure quaternions can be called *canonical*. Given a quaternion q, its three *canonical* involutions are:

$$\begin{array}{lllll} \overline{q}^{\boldsymbol{i}} & = & a+\boldsymbol{i}b-\boldsymbol{j}c-\boldsymbol{k}d & = & -\boldsymbol{i}q\boldsymbol{i} \\ \overline{q}^{\boldsymbol{j}} & = & a-\boldsymbol{i}b+\boldsymbol{j}c-\boldsymbol{k}d & = & -\boldsymbol{j}q\boldsymbol{j} \\ \overline{q}^{\boldsymbol{k}} & = & a-\boldsymbol{i}b-\boldsymbol{j}c+\boldsymbol{k}d & = & -\boldsymbol{k}q\boldsymbol{k} \end{array} \qquad [1.23]$$

Clearly, q and its three canonical involutions allow us to recover the four components a, b, c, d of q by linear combination. Now, the most general definition for involution is:

$$\overline{q}^{\boldsymbol{\mu}} = -\boldsymbol{\mu}q\boldsymbol{\mu} \qquad [1.24]$$

where $\boldsymbol{\mu} \in \mathbf{V}(\mathbb{H})$ and $|\boldsymbol{\mu}| = 1$.

Involutions in $\mathbb{H}$ possess many properties (see [ELL 07d] for details) among which, for $q, p \in \mathbb{H}$ and $\boldsymbol{\mu} \in \mathbf{V}(\mathbb{H})$ and $\boldsymbol{\mu}^2 = -1$ ($\boldsymbol{i}, \boldsymbol{j}$ and $\boldsymbol{k}$ are possible choices for $\boldsymbol{\mu}$):

$$\begin{cases} \overline{qp}^{\boldsymbol{\mu}} & = \overline{q}^{\boldsymbol{\mu}}\overline{p}^{\boldsymbol{\mu}} \\ \overline{\overline{q}^{\boldsymbol{\mu}}}^{\boldsymbol{\mu}} & = q \\ \overline{\overline{q}^{\boldsymbol{i}}}^{\boldsymbol{j}} & = \overline{q}^{\boldsymbol{k}} \end{cases} \qquad [1.25]$$

As $\mathbb{H}$ is a division algebra, any non-null quaternion possesses an inverse. The inverse of a given quaternion $q \in \mathbb{H}$ is given by:

$$q^{-1} = \frac{\overline{q}}{\|q\|} \qquad [1.26]$$

where it can be easily checked that $qq^{-1} = 1$ because of [1.22]. Note that for a pure unit quaternion $\boldsymbol{\mu}$, ($\|\boldsymbol{\mu}\| = 1$ and $\mathrm{S}(\boldsymbol{\mu}) = 0$), the following holds: $\boldsymbol{\mu}^{-1} = -\boldsymbol{\mu}$.

Now that we have introduced the inverse of a quaternion, we are ready to look at the ratio of two quaternions p and q. Ratios must be handled with care in $\mathbb{H}$ (indeed, in any non-commutative algebra), and it is preferable to avoid the p/q notation when possible, as it is ambiguous, since p/q can be interpreted as the product of p by q^{-1};

the notation p/q does not specify the order of the product so that it leaves the possibility for pq^{-1} and $q^{-1}p$. The ambiguity arises from the fact that in general:

$$pq^{-1} \neq q^{-1}p \qquad [1.27]$$

The above non-equality arises from the fact that:

$$pq^{-1} = \frac{p\overline{q}}{\|q\|} \quad \text{while} \quad q^{-1}p = \frac{\overline{q}p}{\|q\|} \qquad [1.28]$$

Thus, it is important to consider ratios as products with the inverse and to take care of the order of the product.

Now, the modulus of a ratio is an interesting quantity to consider, as it does not suffer from the order in the multiplication, due to the property of the modulus given in [1.17]. It thus follows that:

$$\left|q^{-1}p\right| = \left|pq^{-1}\right| = \frac{|p|}{|q|} \qquad [1.29]$$

It can be useful to write a given quaternion $q \in \mathbb{H}$ as a product of a scalar positive number (its modulus) and a unit quaternion. This can be done in the following way:

$$q = |q|\,\tilde{q} = |q|\left(\frac{a}{|q|} + \boldsymbol{i}\frac{b}{|q|} + \boldsymbol{j}\frac{c}{|q|} + \boldsymbol{k}\frac{d}{|q|}\right) \qquad [1.30]$$

where we used the notation $\tilde{q}$ for the *unit modulus version* of q, i.e. $\tilde{q} = q/|q|$ so that $|\tilde{q}| = 1$. Note that the decomposition of q into the product of its modulus and its normalized version is *unique*. The normalized version of q, denoted by $\tilde{q}$, is also sometimes called a *versor*. Finally, it must be emphasized that if q is a *pure* quaternion, i.e. $q \in \mathbf{V}(\mathbb{H})$, then it is uniquely written as $q = |q|\,\boldsymbol{\mu}$ where we have denoted $\tilde{q} = \boldsymbol{\mu}$ to highlight the fact that it is a *pure unit* quaternion.

In [1.5], we introduced the Cartesian form of a quaternion $q \in \mathbb{H}$, in which it is expressed using the sum of a real part $\Re(q)$, an $\boldsymbol{i}$–imaginary part $\Im_{\boldsymbol{i}}(q)$, a $\boldsymbol{j}$–imaginary part $\Im_{\boldsymbol{j}}(q)$ and a $\boldsymbol{k}$–imaginary part $\Im_{\boldsymbol{k}}(q)$. This expression is a special case of the expansion of a quaternion over a 4D basis. The specific basis used in [1.5] is $\{1, \boldsymbol{i}, \boldsymbol{j}, \boldsymbol{k}\}$. This is the *classical* basis used by most authors. Now, it is possible to use a different basis and it turns out that there is an infinite amount of choices for a basis in $\mathbb{H}$. Given two pure unit quaternions $\boldsymbol{\mu}$ and $\boldsymbol{\eta}$, i.e. $\boldsymbol{\mu}^2 = \boldsymbol{\eta}^2 = -1$, which are orthogonal to each other: $\boldsymbol{\mu}\perp\boldsymbol{\eta}$, the set $\{1, \boldsymbol{\mu}, \boldsymbol{\eta}, \boldsymbol{\mu\eta}\}$ is a basis of $\mathbb{H}$. The infinite number of possibilities arises from the infinite number of possible choices for the pair of orthogonal pure unit quaternions $\boldsymbol{\mu}$ and $\boldsymbol{\eta}$. Note that the orthogonality between pure unit quaternions should be understood as:

$$\mathrm{S}(\boldsymbol{\mu}\boldsymbol{\eta}) = \langle \boldsymbol{\mu}, \boldsymbol{\eta} \rangle = 0 \qquad [1.31]$$

which is the scalar part of the quaternion product of two pure quaternions (see the expression for the quaternion product in [1.9]). Also note that the orthogonality condition does not mean that the *quaternion* product between $\boldsymbol{\mu}$ and $\boldsymbol{\eta}$ is equal to zero. On the contrary, as can be seen from [1.9], we have:

$$\boldsymbol{\mu}\boldsymbol{\eta} = \mathbf{V}(\boldsymbol{\mu}\boldsymbol{\eta}) = \boldsymbol{\mu} \times \boldsymbol{\eta} \qquad [1.32]$$

which is indeed a pure unit quaternion orthogonal to both $\boldsymbol{\mu}$ and $\boldsymbol{\eta}$. This makes $\boldsymbol{\mu}\boldsymbol{\eta}$ the third pure unit quaternion that, together with 1, $\boldsymbol{\mu}$ and $\boldsymbol{\eta}$, forms the quaternion basis. Quaternion bases play an important role in the quaternion notation as well as in the computation of quaternion Fourier transforms (QFT) with arbitrary axis, as will be presented in Chapter 3.

The list of quaternion properties presented in this section is not intended to be exhaustive. More properties can be found in various references [CON 03, KAN 89, WAR 97, KUI 02, HAN 06].

1.3. Exponential and logarithm of a quaternion

Among the functions with quaternion-valued arguments that will be considered in the sequel, two will be of major importance: the exponential and logarithm functions. The former will be central to the study and use of QFTs in subsequent chapters.

1.3.1. *Exponential of a pure quaternion*

The exponential function $\exp : \mathbf{V}(\mathbb{H}) \to \mathbb{H}$ can be defined through its power series expansion; thus, for a given (non-null) pure quaternion $\xi \in \mathbf{V}(\mathbb{H})$ written as $\xi = |\xi|\,\boldsymbol{\xi}$ with $\boldsymbol{\xi}$ a pure *unit* quaternion[3] (i.e. $\boldsymbol{\xi} \in \mathbf{V}(\mathbb{H})$ and $\boldsymbol{\xi}^2 = -1$) and $|\xi| \in \mathbb{R}^+$, we can write:

$$e^{\xi} = \sum_{n=0}^{+\infty} \frac{\xi^n}{n!} = \sum_{n=0}^{+\infty} \frac{|\xi|^n\,\boldsymbol{\xi}^n}{n!} \qquad [1.33]$$

3 Just like any quaternion, a *pure* quaternion $\xi \in \mathbf{V}(\mathbb{H})$ can be uniquely written as $\xi = |\xi|\,\boldsymbol{\xi}$. $\boldsymbol{\xi}$ is *unit* quaternion (in this case, pure) and $|\xi|$ is its modulus.

where we make use of the fact that $|\xi|$ and $\boldsymbol{\xi}$ commute as any quaternion commutes with a real number. Now, just like the familiar complex imaginary number I does, a pure unit quaternion behaves, when it is raised to an integer power of n, as:

$$\boldsymbol{\xi}^n = \begin{cases} (-1)^k & \text{if} \quad n = 2k \\ (-1)^k \boldsymbol{\xi} & \text{if} \quad n = 2k+1 \end{cases} \qquad [1.34]$$

With this in mind, the exponential e^{ξ} simply is:

$$e^{\xi} = \sum_{n=0}^{+\infty} (-1)^k \frac{|\xi|^{2k}}{(2k)!} + \boldsymbol{\xi} \sum_{n=0}^{+\infty} (-1)^k \frac{|\xi|^{2k+1}}{(2k+1)!} \qquad [1.35]$$

In this equation, the right-hand side terms are the series expansions of the classical cos and sin functions. We have established that for a pure quaternion $\xi = |\xi| \boldsymbol{\xi}$, the exponential of ξ is:

$$e^{|\xi|\boldsymbol{\xi}} = \cos|\xi| + \boldsymbol{\xi} \sin|\xi| \qquad [1.36]$$

This shows that the exponential of a pure quaternion can be expressed easily in terms of cosine and sine functions just as in the complex case. The difference lies in the axis $\boldsymbol{\xi}$ which is a unit pure quaternion, while the argument is the *modulus* of ξ. Thus, the exponential of a pure quaternion is a full quaternion, with real/scalar part $\cos|\xi|$ and vector part $\boldsymbol{\xi} \sin|\xi|$.

It must also be noticed that the exponential of a pure quaternion is always of unit modulus:

$$\left|e^{\xi}\right| = 1, \qquad \forall \xi \in \mathbf{V}(\mathbb{H}) \qquad [1.37]$$

which is easily verified using [1.36] as well as property [1.34] with $n = 2$.

Thus, we conclude that the exponential of a pure quaternion is a full unit quaternion. Now, the reciprocal property is interesting: any full unit quaternion can be expressed as the exponential of a pure quaternion[4]. This is a remarkable property that will be tackled in section 1.3.3 for the expression of the logarithm of a quaternion, and especially for the polar form and Euler formula over $\mathbb{H}$ (sections 1.4.1 and 1.4.1.1).

4 A full unit quaternion q is *uniquely* expressed as the exponential of a pure quaternion, i.e. $q = \exp(\xi)$ with $\xi = |\xi| \boldsymbol{\xi}$, iff $|\xi| \in [0, 2\pi[$.

Another important property, which differs drastically from the complex exponential case, is that the product of two exponentials of pure quaternions[5] is not an exponential with argument equal to the sum of the arguments of the original exponentials. This means that for $\alpha, \beta \in \mathbb{R}^+$ and any two distinct pure unit quaternions $\boldsymbol{\mu}, \boldsymbol{\nu}$, we have:

$$e^{\boldsymbol{\nu}\alpha} e^{\boldsymbol{\mu}\beta} \neq e^{\boldsymbol{\nu}\alpha + \boldsymbol{\mu}\beta} \quad [1.38]$$

One should note that in the special case where $\boldsymbol{\mu} = \boldsymbol{\nu}$, the equality is fulfilled, meaning that for $\alpha, \beta \in \mathbb{R}^+$ and $\boldsymbol{\mu}$ a pure unit quaternion, one has $\exp(\boldsymbol{\mu}\alpha)\exp(\boldsymbol{\mu}\beta) = \exp(\boldsymbol{\mu}(\alpha + \beta))$. In fact, this is a well-known property of exponential functions over non-commutative algebras (for example, the exponentials of matrices) and a consequence of the Baker–Campbell–Hausdorff formula (see, for example, [GIL 08] for illustrations of this formula).

The exponential function of a pure quaternion will be of use when considering polar representations of quaternions as well as when defining QFTs.

1.3.2. *Exponential of a full quaternion*

We have already introduced the exponential of a pure quaternion in section 1.3.1 for use later in the definition of Fourier transforms. Here, we consider the exponential of *full* quaternions, i.e. the function $\exp : \mathbb{H} \to \mathbb{H}$. Just as in section 1.3.1, the exponential function is directly defined through its series expansion that is absolutely convergent. Thus, for a quaternion $q \in \mathbb{H}$, its exponential is given by:

$$e^q = \sum_{n=0}^{+\infty} \frac{q^n}{n!} \quad [1.39]$$

Now, recalling that any quaternion can be written as $q = \mathrm{S}(q) + \mathbf{V}(q)$, it follows directly that:

$$e^q = e^{\mathrm{S}(q)} e^{\mathbf{V}(q)} \quad [1.40]$$

leading to the special case of the exponential of a pure quaternion if $\mathrm{S}(q) = 0$. Now, it is also possible to expand the $e^{\mathbf{V}(q)}$ part into cos and sin, as $\mathbf{V}(q) \in \mathbf{V}(\mathbb{H})$, leading to the following expression for the exponential of q:

$$e^q = e^{\mathrm{S}(q)} \left(\cos|\mathbf{V}(q)| + \widetilde{\mathbf{V}(q)} \sin|\mathbf{V}(q)| \right) \quad [1.41]$$

5 This generalizes to the exponential of full quaternions.

with $\mathbf{V}(q) = |\mathbf{V}(q)|\,\widetilde{\mathbf{V}(q)}$, where $\widetilde{\mathbf{V}(q)}$ is the normalized vector part of q, i.e. the versor associated with $\mathbf{V}(q)$. Note that due to the fact that $e^{\mathrm{S}(q)}$ is real-valued, it commutes with the exponential of the vector part $\mathbf{V}(q)$. Finally, note that given a quaternion $q \in \mathbb{H}$ and a complex number $z \in \mathbb{C}$, their exponentials e^q and e^z are isomorphic, provided that $|q| = |z|$ and $\mathrm{S}(q)\,/\,|\mathbf{V}(q)| = \Re(z)/\Im(z)$.

1.3.3. *Logarithm of a quaternion*

The logarithm of a quaternion q is simply defined as the inverse of the exponential function, so that for $q, p \in \mathbb{H}$, we have:

$$p = \ln q \qquad [1.42]$$

which means that $e^p = q$. However, it is possible to obtain an expression for the logarithm of q in terms of its elements. First, we recall that $q \in \mathbb{H}$ can be expressed as $q = |q|\,\tilde{q}$, where $\tilde{q}$ is the normalized version of q. Now, it follows directly that the logarithm of q is:

$$\ln q = \ln|q| + \ln\tilde{q} \qquad [1.43]$$

Now, the second term on the right-hand side of the equation can be found to have an interesting literal expression. There is no doubt that $\ln|q| \in \mathbb{R}$; now, we are going to show that $\ln\widetilde{q}$ is a pure quaternion. First, remember from section 1.3.1 that, as $\widetilde{q}$ is a full unit quaternion, it can be expressed as the exponential of a pure (non-unit) quaternion ξ :

$$\tilde{q} = e^{\xi} = e^{|\xi|\boldsymbol{\xi}} = e^{\phi_q \boldsymbol{\mu}_q} \qquad [1.44]$$

where we used the following notation:

$$\begin{cases} \phi_q = |\xi| & \in \mathbb{R}^+ \\ \boldsymbol{\mu}_q = \boldsymbol{\xi} & \in \mathbf{V}(\mathbb{H}) \end{cases} \qquad [1.45]$$

This notation was chosen as ϕ_q and $\boldsymbol{\mu}_q$ will be identified later in section 1.4.

Now, with this notation, and by direct substitution of [1.44] into [1.43], we have:

$$\ln q = \ln|q| + \boldsymbol{\mu}_q\phi_q \qquad [1.46]$$

where we use the fact that $\boldsymbol{\mu}_q\phi_q = \phi_q\boldsymbol{\mu}_q$ as ϕ_q is real-valued. This expression of the logarithm of a quaternion can be seen as a generalization of the well-known logarithm of a complex number. In particular, the fact that we fixed the range of values to be taken by the argument ϕ_q between 0 and 2π ensures that the logarithm function with a quaternion argument is not multivalued.

1.4. Representations

Here we describe several existing representations of quaternions that will be used throughout the book.

1.4.1. *Polar forms*

1.4.1.1. *Euler formula*

In addition to the Cartesian form of q given in [1.1], there are several other representations for quaternions that have been introduced since their discovery. One of the most important notations, which was introduced by Hamilton himself and is called the *polar form* of a quaternion q, is the quaternion equivalent of what Richard Feynman called the "jewel" of complex numbers, namely the Euler formula. It encapsulates the link between geometry and algebra. For a quaternion $q \in \mathbb{H}$, it reads as:

$$q = |q|\, e^{\boldsymbol{\mu}_q \phi_q} = |q| \left(\cos\phi_q + \boldsymbol{\mu}_q \sin\phi_q\right) \qquad [1.47]$$

where $|q|$ is the *modulus*, $\boldsymbol{\mu}_q$ is called the *axis* and ϕ_q is the *phase/argument* of q. The elements of the polar form are given in terms of the Cartesian components as:

$$\begin{cases} |q| & = \sqrt{a^2 + b^2 + c^2 + d^2} \\ \boldsymbol{\mu}_q & = \dfrac{b\boldsymbol{i} + c\boldsymbol{j} + d\boldsymbol{k}}{\sqrt{b^2 + c^2 + d^2}} \\ \phi_q & = \arctan\left(\dfrac{\sqrt{b^2 + c^2 + d^2}}{a}\right) \end{cases} \qquad [1.48]$$

where $\boldsymbol{\mu}_q \in \mathbf{V}(\mathbb{H})$ and $|\boldsymbol{\mu}_q| = 1$. From this notation, it follows that unit quaternions have a modulus equal to 1, and pure quaternions (i.e. with $a = 0$) have a phase equal to $\pi/2$. The polar form allows us to link geometrical concepts such as rotations in 3D and 4D to the algebra of quaternions. This will be elaborated upon in Chapter 2. It is interesting to look at the special case of quaternions with unit modulus. In such a case, the Euler form simply reads as:

$$e^{\boldsymbol{\mu}\phi} = \cos\phi + \boldsymbol{\mu}\sin\phi \qquad [1.49]$$

still with $\boldsymbol{\mu} \in \mathbf{V}(H)$ and $|\boldsymbol{\mu}| = 1$. Note that $\cos\phi \in \mathbb{R}$ and $\boldsymbol{\mu}\sin\phi$ is the vector part of this quaternion. As a result, taking its conjugate consists of simply negating the vector part, giving the following identity:

$$\cos\phi - \boldsymbol{\mu}\sin\phi = e^{-\boldsymbol{\mu}\phi} \qquad [1.50]$$

where one can see that conjugating a unit quaternion consists of either reversing its axis or negating its phase. Note that this could have been guessed from the fact that for unit quaternions $q \in \mathbb{H}$, i.e. $|q| = 1$, then $\overline{q} = q^{-1}$. Now, due to expressions [1.49] and [1.50], we can express the sin and cos functions in the following way:

$$\begin{cases} \cos\phi &= \frac{1}{2}\left(e^{\boldsymbol{\mu}\phi} + e^{-\boldsymbol{\mu}\phi}\right) \\ \sin\phi &= -\frac{1}{2}\boldsymbol{\mu}\left(e^{\boldsymbol{\mu}\phi} - e^{-\boldsymbol{\mu}\phi}\right) \end{cases} \qquad [1.51]$$

which will prove to be useful later in the study of QFTs in Chapter 3.

1.4.1.2. *The Euler angle parameterization polar form*

Another *polar* form that was used in [BÜL 01] has a direct connection to Euler angles [ALT 86]. As there are 12 possible conventions when using Euler angles, there are equivalently 12 possible polar forms. Here, we illustrate the polar form with the XYZ convention as used in [BÜL 01]. With this convention, a quaternion $q \in \mathbb{H}$ can be expressed as:

$$q = |q|\, e^{\eta \boldsymbol{i}} e^{\kappa \boldsymbol{j}} e^{\phi \boldsymbol{k}} \qquad [1.52]$$

The three angles of q, i.e. $\eta \in [0, 2\pi)$, $\kappa \in [0, \pi)$, $\phi \in [0, 2\pi)$, can be identified as three *phases*, which are related to Euler angles when $|q| = 1$.

1.4.1.3. *The Cayley–Dickson form*

It can also be useful to consider a quaternion as a pair of complex numbers with specific multiplication rules. This is the idea behind the Cayley–Dickson (CD) form of a quaternion q, which can be obtained using the doubling procedure [KAN 89]. Thus, a quaternion $q \in \mathbb{H}$ has a CD form that reads:

$$q = z_1 + z_2 \boldsymbol{j} \qquad [1.53]$$

where $z_1 = a + \boldsymbol{i}b \in \mathbb{C}_{\boldsymbol{i}}$ and $z_2 = c + \boldsymbol{i}d \in \mathbb{C}_{\boldsymbol{i}}$. Obviously, there is an infinite number of CD forms, depending on the unit pure quaternion that is used to split the quaternion into two different planes. This will be detailed in section 1.4.1.4. All the quaternion properties could be rephrased by this notation. However, we will not do so as it is not of use in the following. As an example, we give the expression of the quaternion conjugation in the CD notation, which is:

$$\overline{q} = z_1^{\star} - z_2 \boldsymbol{j} \qquad [1.54]$$

where * denotes the classical complex conjugation.

1.4.1.4. *The ortho-split or symplectic form*

A more general representation exists, in the spirit of the CD form, that allows the interpretation of a quaternion in terms of two complex numbers found in two non-intersecting two-dimensional (2D) planes in 4D space[6]. Consider a basis in $\mathbb{H}$: $\{1, \boldsymbol{\mu}, \boldsymbol{\nu}, \boldsymbol{\mu\nu}\}$, where $\boldsymbol{\mu}^2 = \boldsymbol{\nu}^2 = -1$ and $\mathrm{S}(\boldsymbol{\mu\nu}) = 0$, i.e. $\boldsymbol{\mu} \perp \boldsymbol{\nu}$. $\boldsymbol{\mu}$ and $\boldsymbol{\nu}$ are pure unit quaternions (square roots of -1). As a result, $\boldsymbol{\mu\nu} \perp \boldsymbol{\mu}$ and $\boldsymbol{\mu\nu} \perp \boldsymbol{\nu}$. Then, any quaternion $q \in \mathbb{H}$ can be written as:

$$q = (a' + b'\boldsymbol{\mu}) + (c' + d'\boldsymbol{\mu})\boldsymbol{\nu} \qquad [1.55]$$

where the two above-mentioned planes are spanned by $\{1, \boldsymbol{\mu}\}$ and $\{\boldsymbol{\nu}, \boldsymbol{\mu\nu}\}$. This representation of a quaternion is sometimes referred to as its *ortho-split* representation or *symplectic* form[7] as in [ELL 07c]. It is also known as the decomposition of q into its *simplex* part and *perplex* part. Using the quantities defined in [1.55] the *simplex* part of q is $(a' + b'\boldsymbol{\mu})$ and the *perplex* part of q is $(c' + d'\boldsymbol{\mu})$. Later in this book, we will sometimes make use of the notation:

$$q = q_s + q_p\boldsymbol{\nu} \qquad [1.56]$$

with notations $q_s \in \mathbb{C}_{\boldsymbol{\mu}}$ for the simplex part of $q \in \mathbb{H}$ and $q_p \in \mathbb{C}_{\boldsymbol{\mu}}$ for its perplex part, both understood with respect to the axis $\boldsymbol{\nu}$ satisfying $\boldsymbol{\mu} \perp \boldsymbol{\nu}$. Using this symplectic notation, a quaternion q can be written as:

$$q = q_+ + q_- \qquad [1.57]$$

where:

$$\begin{cases} q_+ &= \frac{1}{2}\left(q + \boldsymbol{\mu}\, q\, \boldsymbol{\nu}\right) \\ q_- &= \frac{1}{2}\left(q - \boldsymbol{\mu}\, q\, \boldsymbol{\nu}\right) \end{cases} \qquad [1.58]$$

which can be seen as a generalization of the expressions in [1.19] using a quaternion q and its conjugate.

We also mention here the *swap-rule* for the symplectic notation, which will be of use in the even-odd split study of the QFT in section 3.1.3. If, in a symplectic

6 The two planes are called non-intersecting even though it is not strictly true as they intersect at the origin. However, as this is the only point in 4D space where the planes intersect, we will keep the terminology "non-intersecting".

7 Note that the terminology *symplectic* used here is different from the classically used term "symplectic" used in differential geometry, topology or group theory where it designates a special non-degenerate 2-form.

decomposition, one needs to have the imaginary axis $\boldsymbol{\nu}$ to be placed on the left, then one can simply write q as:

$$q = (a' + b'\boldsymbol{\mu}) + (c' - d'\boldsymbol{\mu})\boldsymbol{\nu} = q_s + \boldsymbol{\nu} q_p^\star \qquad [1.59]$$

where the complex conjugation * is used on q_p despite its quaternionic nature as it is isomorphic to a complex number, i.e. $q_p \in \mathbb{C}_{\boldsymbol{\mu}}$.

Note finally that choosing a basis in $\mathbb{H}$ to express a quaternion q induces a choice of symplectic representation. The symplectic form is thus a generalization of the CD form presented in section 1.4.1.3.

1.4.1.5. *The polar Cayley–Dickson form*

Recently, a new representation was introduced for elements of $\mathbb{H}$, called the *polar Cayley–Dickson* representation [SAN 10]. Given a quaternion $q \in \mathbb{H}$, its polar CD form is:

$$q = \rho_q e^{\Phi_q \boldsymbol{j}} \qquad [1.60]$$

with $\rho_q \in \mathbb{C}$ the *complex modulus* of q and $\Phi_q \in \mathbb{C}$ its *complex phase*. This representation of a quaternion is unique. Its construction is given in detail in [SAN 10]. Here, we present the main expressions of the polar CD form, without details of the sign ambiguity that is known to exist during the construction of the polar CD form of a quaternion from its Cartesian form. For a complete discussion of this sign issue, one should refer to [SAN 10]. Now, given a quaternion $q \in \mathbb{H}$, with CD form $q = z_1 + z_2\boldsymbol{j}$, its complex modulus and phase ρ_q and Φ_q are obtained by:

$$\begin{cases} \rho_q & = z_1 \dfrac{|q|}{|z_1|} \\ \Phi_q & = -\ln\left(\dfrac{z_1^\star q}{\|z_1\|}\boldsymbol{j}\right) \end{cases} \qquad [1.61]$$

Note that in the expression for Φ_q, the argument of the logarithm is a product between a complex number and a quaternion, meaning that the order is important and that q should be multiplied from the left by $z_1^\star$. Note also that the product $z_1^\star q$ is rather special in that if we replace q by its CD form, we get $z_1^\star q = |z_1|^2 + z_1^\star z_2 \boldsymbol{j}$, which is a *degenerate* quaternion having a real part, an $\Im_{\boldsymbol{j}}$ part and an $\Im_{\boldsymbol{k}}$ part, but its $\Im_{\boldsymbol{i}}$ part is null. As a result, taking the logarithm of such a degenerate quaternion leads to a quaternion with only a $\Im_{\boldsymbol{j}}$ and a $\Im_{\boldsymbol{k}}$ part. The negation and right multiplication by $\boldsymbol{j}$ finally lead to the fact that Φ_q is a complex number from $\mathbb{C}_{\boldsymbol{i}}$. It is also interesting to note that any ortho-split form as introduced in section 1.4.1.3 will lead to another possible polar CD form based on the axis used for the split.

An illustration of the usefulness of the polar CD form is made in section 4.3.4.

1.4.2. *The $\mathbb{C}_j$-pair notation*

In the style of the *ortho-split* notation or CD form, it is possible to think of a quaternion as an *ordered pair of complex numbers* and to manipulate them as such. Here, we mention this notation, called the $\mathbb{C}_j$-pair associated with a quaternion, as it highlights a very special product, namely the *bicomplex product* of two quaternions. This product will be of use when studying convolution for the one-dimensional (1D) QFT in Chapter 4.

Consider the quaternion $q = a + \boldsymbol{i}b + \boldsymbol{j}c + \boldsymbol{k}d \in \mathbb{H}$ and let us write it as $q = q_1 + \boldsymbol{i}q_2$. In this way, q can be considered as the pair $q = (q_1, q_2)$ with $q_1 = a + \boldsymbol{j}c$ and $q_2 = b + \boldsymbol{j}d$ being elements of $\mathbb{C}_{\boldsymbol{j}}$. Obviously, it is a special case of an ortho-split decomposition, with axis $\boldsymbol{i}$, where the quaternion q is simply understood as a pair of complex numbers.

Now, with this notation, we could rewrite anything related to quaternion algebra. For example, the conjugate of q is $\overline{q} = (q_1^\star, -q_2)$. More interestingly, if we consider the two quaternions $q = (q_1, q_2)$ and $p = (p_1, p_2)$, then their (quaternion) product is:

$$qp = (q_1p_1 - q_2^\star p_2\, ,\, q_2p_1 + q_1^\star p_2) \qquad [1.62]$$

and the inverse of q is simply:

$$q^{-1} = \left(\frac{q_1^\star}{|q|^2}\, ,\, \frac{-q_2}{|q|^2} \right) \qquad [1.63]$$

Now, the involutions of q with respect to the canonical axis of $\mathbb{H}$ can be expressed as:

$$\begin{cases} \overline{q}^{\boldsymbol{i}} &= (q_1\, ,\, -q_2) \\ \overline{q}^{\boldsymbol{j}} &= (q_1^\star\, ,\; \; q_2^\star) \\ \overline{q}^{\boldsymbol{k}} &= (q_1^\star\, ,\, -q_2^\star) \end{cases} \qquad [1.64]$$

Indeed, all the quaternion operations can be handled with this notation, but we will make special use in the following of one particular operation that comes in a very useful form when using the complex pair notation: the *bicomplex product* denoted by $\odot$. Given two quaternions in their complex pair forms $q = (q_1, q_2), p = (p_1, p_2) \in \mathbb{H}$, their *bicomplex product* is:

$$q \odot p = (q_1p_1 - q_2p_2\, ,\, q_2p_1 + q_1p_2) \qquad [1.65]$$

where we can see the difference with the standard quaternion product in [1.62]. A very remarkable property of the *bicomplex product* of two quaternions is that it is *commutative*. This can be easily seen from the expression of $p \odot q$:

$$p \odot q = (p_1 q_1 - p_2 q_2 \, , \, p_2 q_1 + p_1 q_2) \qquad [1.66]$$

and by remembering that $p_{1,2}$ and $q_{1,2}$ are complex numbers and thus $p_i q_j = q_j p_i$ for any pair $i, j = 1, 2$. This leads to the following interesting property:

$$q \odot p = p \odot q \qquad \forall p, q \in \mathbb{H} \qquad [1.67]$$

Note that the commutativity of this product is associated with the $\mathbb{C}_{\boldsymbol{j}}$-pairing notation. If we use a different way to divide a quaternion into two complex numbers (a $\mathbb{C}_{\boldsymbol{\mu}}$ pairing with $\boldsymbol{\mu}$ a square root of -1), then we need to carefully look at how to define an associated commutative product.

Rather than providing a lengthy study of the commutative product, we give here some properties of this special product over $\mathbb{H}$, for any quaternion $q = (q_1, q_2) = (a + \boldsymbol{j}c, b + \boldsymbol{j}d)$, any $\mathbb{C}_{\boldsymbol{i}}$ number $z = \Re(z) + \boldsymbol{i}\Im_{\boldsymbol{i}}(z) = (\Re(z), \Im_{\boldsymbol{i}}(z))$, any $\mathbb{C}_{\boldsymbol{j}}$ number $w = \Re(w) + \boldsymbol{j}\Im_{\boldsymbol{j}}(w) = (w, 0)$ and $\mathbb{C}_{\boldsymbol{k}}$ number $s = \Re(s) + \boldsymbol{k}\Im_{\boldsymbol{k}}(s) = (\Re(s), \boldsymbol{j}\Im_{\boldsymbol{k}}(s))$. Then the following properties hold:

$$\begin{cases} q \odot w & = (q_1, q_2) \odot (w, 0) = (q_1 w \, , \, q_2 w) \\ q \odot z & = (q_1, q_2) \odot (\Re(z), \, \Im_{\boldsymbol{i}}(z)) \\ & = (q_1 \Re(z) - q_2 \Im_{\boldsymbol{i}}(z) \, , \, q_2 \Re(z) + q_1 \Im_{\boldsymbol{i}}(z)) \\ q \odot s & = (q_1, q_2) \odot (\Re(s), \, \boldsymbol{j}\Im_{\boldsymbol{k}}(s)) \\ & = (q_1 \Re(s) - \boldsymbol{j} q_2 \Im_{\boldsymbol{k}}(s) \, , \, q_2 \Re(s) + \boldsymbol{j} q_1 \Im_{\boldsymbol{k}}(s)) \\ q \odot \boldsymbol{i} & = (q_1, q_2) \odot (0, 1) = (-q_2, q_1) \\ q \odot \boldsymbol{j} & = (q_1, q_2) \odot (\boldsymbol{j}, 0) = (q_1 \boldsymbol{j}, q_2 \boldsymbol{j}) \\ q \odot \boldsymbol{k} & = (q_1, q_2) \odot (0, \boldsymbol{j}) = (-q_2 \boldsymbol{j}, q_1 \boldsymbol{j}) \\ \boldsymbol{i} \odot \boldsymbol{j} & = \boldsymbol{j} \odot \boldsymbol{i} = \boldsymbol{k} \\ \boldsymbol{k} \odot \boldsymbol{j} & = \boldsymbol{j} \odot \boldsymbol{k} = -\boldsymbol{i} \\ \boldsymbol{i} \odot \boldsymbol{k} & = \boldsymbol{k} \odot \boldsymbol{i} = -\boldsymbol{j} \\ \boldsymbol{i} \odot \boldsymbol{i} & = -1 \\ \boldsymbol{j} \odot \boldsymbol{j} & = -1 \\ \boldsymbol{k} \odot \boldsymbol{k} & = 1 \end{cases} \qquad [1.68]$$

where all the given results can be checked by direct calculation using the bicomplex product expression given in [1.65].

As previously mentioned, the *bicomplex product* will not be encountered many times in this book, but it will be of use in section 4.3 when studying the convolution theorem for the QFT of complex-valued signals. The interested reader is referred to [PRI 91, ROC 04] for more materials on bicomplex numbers.

1.4.3. $\mathbb{R}$ *and* $\mathbb{C}$ *matrix representations*

In addition to scalar representations, there exist two matrix representations over $\mathbb{R}$ and $\mathbb{C}$ that are isomorphic to $\mathbb{H}$. Given a quaternion $q \in \mathbb{H}$, its real matrix representation, denoted as $\mathcal{M}_{\mathbb{R}}(q) \in \mathbb{R}^{4\times 4}$, is given by[8]:

$$\begin{aligned}
\mathcal{M}_{\mathbb{R}}(q) &= \begin{bmatrix} a & b & c & d \\ -b & a & -d & c \\ -c & d & a & -b \\ -d & -c & b & a \end{bmatrix} \\
&= a\begin{bmatrix} 1 & 0 & 0 & 0 \\ 0 & 1 & 0 & 0 \\ 0 & 0 & 1 & 0 \\ 0 & 0 & 0 & 1 \end{bmatrix} + b\begin{bmatrix} 0 & 1 & 0 & 0 \\ -1 & 0 & 0 & 0 \\ 0 & 0 & 0 & -1 \\ 0 & 0 & 1 & 0 \end{bmatrix} \\
&\quad + c\begin{bmatrix} 0 & 0 & 1 & 0 \\ 0 & 0 & 0 & 1 \\ -1 & 0 & 0 & 0 \\ 0 & -1 & 0 & 0 \end{bmatrix} + d\begin{bmatrix} 0 & 0 & 0 & 1 \\ 0 & 0 & -1 & 0 \\ 0 & 1 & 0 & 0 \\ -1 & 0 & 0 & 0 \end{bmatrix}
\end{aligned} \quad [1.69]$$

and for any quaternion $q \in \mathbb{H}$, its complex matrix representation, denoted as $\mathcal{M}_{\mathbb{C}}(q) \in \mathbb{C}^{2\times 2}$, is of the form:

$$\begin{aligned}
\mathcal{M}_{\mathbb{C}}(q) &= \begin{bmatrix} z_1 & -z_2^\star \\ z_2 & z_1^\star \end{bmatrix} \\
&= a\begin{bmatrix} 1 & 0 \\ 0 & 1 \end{bmatrix} + b\begin{bmatrix} i & 0 \\ 0 & -i \end{bmatrix} + c\begin{bmatrix} 0 & -1 \\ 1 & 0 \end{bmatrix} + d\begin{bmatrix} 0 & -i \\ i & 0 \end{bmatrix}
\end{aligned} \quad [1.70]$$

Comparing the real and complex matrix expressions for q with the Cartesian form, one gets a direct identification of the imaginary units over $\mathbb{H}$, i.e. $\boldsymbol{i}, \boldsymbol{j}$ and $\boldsymbol{k}$ with their real 4×4 and complex 2×2 matrix forms. As a real vector space, quaternions are spanned by these four real or complex matrices. For both of these matrix representations, the quaternion product is identified with the matrix product over $\mathbb{R}^{4\times 4}$ and $\mathbb{C}^{2\times 2}$. From a practical point of view, it is thus possible to manipulate

8 The transpose of the matrix representation given here is also valid.

quaternions through their matrix representations. However, as noted in section I.6.2, this is not optimal from a computational point of view as it requires much more computation and storage than necessary. Matrix representations of quaternions have been found to be of interest in the study of matrices with quaternion entries (for example [ZHA 97]), and in many more application fields.

1.5. Powers of a quaternion

In many situations, we need to multiply quaternions together. Due to the fact that $\mathbb{H}$ is an algebra, the product of two quaternions is still a quaternion. Now, if we multiply a quaternion $q \in \mathbb{H}$, say n times, by itself, it is interesting to look at how the components of the *n*th power of q can be expressed using the components of q. First, note that a quaternion commutes with itself. This may seem trivial, but it is interesting to note that the right and left multiplications are equivalent in this special case. Now, assuming that q has the polar form given in [1.47], it follows directly that:

$$q^n = |q|^n \, e^{n\boldsymbol{\mu}_q \phi_q} = |q|^n \left(\cos(n\phi_q) + \boldsymbol{\mu}_q \sin(n\phi_q)\right) \qquad [1.71]$$

This is simply a generalization of the de Moivre formula for quaternions [KAN 89]. The demonstration can be carried out easily by using the series expansion of the exponential and the property of the *n*th power of the pure unit imaginary $\boldsymbol{\mu}_q$ (see [1.34]). Apart from its simplicity, this formula has useful consequences in the use of quaternions to represent rotations. Anticipating results from Chapter 2, multiplying[9] n times by the same quaternion q will be shown to be equivalent to performing a rotation around the axis $\boldsymbol{\mu}_q$ by an angle $n\phi_q/2$.

1.6. Subfields

As previously mentioned, quaternions form a 4D algebra over $\mathbb{R}$. A quaternion q with no vector part, $\mathbf{V}(q) = 0$, is thus a real number, i.e. $q \in \mathbb{R}$ if $\mathbf{V}(q) = 0$. Similarly, any quaternion with two of its three imaginary components equal to zero is homomorphic to a complex number. This means that for $q \in \mathbb{H}$ of the form:

$$q = a + w\boldsymbol{\mu} \qquad [1.72]$$

with $\boldsymbol{\mu} = \boldsymbol{i}, \boldsymbol{j}, \boldsymbol{k}$, q is equivalent to a complex number. This is to say that numbers taking values in $\mathbb{R} \oplus \boldsymbol{\mu}\mathbb{R}$, with $\boldsymbol{\mu}^2 = -1$, form an algebra homomorphic to complex numbers. q is said to be an element of $\mathbb{C}_{\boldsymbol{\mu}}$. Obviously, we can see that such a construction is the restriction of $\mathbb{H}$ to a 2D plane spanned by $\{1, \boldsymbol{\mu}\}$, just as the

9 The exact way to perform rotations using quaternion multiplication is presented in detail in sections 2.1.2 and 2.1.6.

complex plane is classically known to be spanned by $\{1, I\}$. Now, the above-mentioned $\boldsymbol{\mu}$ was required to be one of the three "classical" imaginary units from $\mathbb{H}$. In fact, it may be any root of -1 in $\mathbb{H}$ so that the subfield (equivalent to the complex field) is now a different 2D plane inside the 4D space of $\mathbb{H}$. Obviously, there are infinitely numerous such subspaces that can be defined within $\mathbb{H}$. Arbitrarily choosing one such subfield yields a new basis for the expression of any quaternion $q \in \mathbb{H}$ through the CD form given in section 1.4.1.4. This shows also that $\mathbb{H}$ contains two copies of $\mathbb{C}$ as subfields.

2

Geometric Applications

The multiplication rules for the quaternion operators $\boldsymbol{i}, \boldsymbol{j}$ and $\boldsymbol{k}$ provide a double meaning to quaternion numbers: on the one hand, they are geometric forms (e.g.vectors in four-dimensional (4D) space) and, on the other hand, they are transformation operators (e.g. rotation operators in three-dimensional (3D) space). Their usefulness as a tool in solving real-world problems is related to the various thought processes, or mental models, used to describe how they work. The geometric forms of signal and image samples upon which the models are based are inherently linked into the quaternion Fourier transform of those samples. There are three models: 3D and 4D vectors, spherical geometry and 3D projective space.

2.1. Euclidean geometry (3D and 4D)

The ability of quaternions to represent geometric transformations has been known right from their discovery and it has been studied and applied in many areas [ALT 86, CON 03, HAN 06, KUI 02, WAR 97, SHO 85]. It is standard knowledge that 3D geometry can be well-handled using quaternions. However, 4D geometry is also a major concept that is inherent to quaternions. A very good reference on 4D geometry using quaternions is Coxeter's paper [COX 46]. Material in the following sections can be found in [COX 46] together with [ELL 07a]. We introduce here the major geometrical transformations that will be of use in the later chapters.

When considering 3D and 4D transformations, it will be assumed that full quaternions (i.e. elements of $\mathbb{H}$) represent points or vectors in 4D space and pure quaternions (i.e. elements of $\mathbf{V}(\mathbb{H})$) represent points or vectors in 3D space.

2.1.1. *3D reflections*

Given a 3D vector $\mathbf{p}$, represented as a pure quaternion, the reflection of $\mathbf{p}$ across the plane defined by its normal vector $\boldsymbol{\eta}$ (i.e. $\boldsymbol{\eta} \in \mathbf{V}(\mathbb{H})$ and $|\boldsymbol{\eta}| = 1$ so that it is a *pure unit quaternion*) is given by:

$$\mathcal{P}_{\boldsymbol{\eta}}[\mathbf{p}] = \boldsymbol{\eta}\mathbf{p}\boldsymbol{\eta} \qquad [2.1]$$

Reflections are the simplest isometric transformations and the basic bricks for building up more elaborate isometric transformations, as is well known (see for example [COX 74, chapter III]). Finally, it is interesting to note that reflections are their own inverse, meaning that if the same reflection is applied twice, one returns to the original vector:

$$\mathcal{P}_{\boldsymbol{\eta}}[\mathcal{P}_{\boldsymbol{\eta}}[\mathbf{p}]] = \boldsymbol{\eta}\,(\boldsymbol{\eta}\mathbf{p}\boldsymbol{\eta})\,\boldsymbol{\eta} = \boldsymbol{\eta}^2\mathbf{p}\,\boldsymbol{\eta}^2 = \mathbf{p} \qquad [2.2]$$

which is true due to the fact that $\boldsymbol{\eta}$ is a pure unit quaternion, i.e. $\boldsymbol{\eta}^2 = -1$.

2.1.2. *3D rotations*

Given a 3D vector $\mathbf{p}$, represented as a pure quaternion, the rotations of $\mathbf{p}$ in 3D space are encoded using one unit quaternion q (i.e. $q \in \mathbb{H}$ and $|q| = 1$) and are given by:

$$\mathcal{R}_q[\mathbf{p}] = q\mathbf{p}\overline{q} \qquad [2.3]$$

where the operator $\mathcal{R}_q[.]$ denotes the rotation with respect to q. As q is a unit quaternion, it can be expressed in its polar (Euler) form as $q = e^{\boldsymbol{\mu}_q\phi_q}$, with $\boldsymbol{\mu}_q$ its axis and ϕ_q its angle/phase, as explained in section 1.4.1.1.

The rotation operator $\mathcal{R}_q[.]$ consists of a rotation around the axis $\boldsymbol{\mu}_q$, with an angle of $2\phi_q$ (twice the angle of q).

Note that, as q is a pure quaternion, then $\overline{q} = q^{-1}$, allowing the alternate notation for the rotation operator: $\mathcal{R}_q[\mathbf{p}] = q\mathbf{p}q^{-1}$.

It is noticeable that the inverse rotation operator $\mathcal{R}_q^{-1}[.]$ is given by:

$$\mathcal{R}_q^{-1}[\mathbf{p}] = \overline{q}\mathbf{p}q \qquad [2.4]$$

which is equivalent to a direct rotation encoded with $\overline{q} = q^{-1}$. This means that the axis is the same but that the angle is exactly the opposite of the one in the rotation $\mathcal{R}_q[.]$.

A very good property of quaternions with respect to rotations is that, when composing such transformations, the resulting transformation can be expressed in terms of the product of the quaternions that represent the individual rotations. Indeed, the composition of two rotations, first $\mathcal{R}_{q_1}[.]$ and then $\mathcal{R}_{q_2}[.]$, applied to a vector p takes the form:

$$\mathcal{R}_{q_2}[\mathcal{R}_{q_1}[\mathbf{p}]] = q_2\left(q_1\mathbf{p}\overline{q_1}\right)\overline{q_2} = q_2q_1\mathbf{p}\overline{(q_2q_1)} \qquad [2.5]$$

which is equivalent to a rotation operator $\mathcal{R}_{q_2q_1}[.]$ encoded using the quaternion q_2q_1. Note that q_2q_1 is a unit full quaternion as it is the product of two unit full quaternions. Obviously, the composition of rotations can be handled easily due to the fact that the group of unit quaternions (isomorphic to $SU(2)$) is the double universal covering of the 3D space rotation group $SO(3)$ [ALT 86].

Generalizing the idea of composition, we can see that composing an arbitrary number of rotations using n different quaternions $q_1, q_2, \ldots, q_n$ is equivalent to performing one rotation with the operator $\mathcal{R}_{q_nq_{n-1}\ldots q_2q_1}[.]$. A special case of composition is applying the same rotation n times. Applying the operator $\mathcal{R}_q[.]$ n times will look like:

$$\underbrace{\mathcal{R}_q\left[\mathcal{R}_q\left[\ldots\, \mathcal{R}_q\left[\mathcal{R}_q\left[\mathbf{p}\right]\right]\, \ldots\right]\right]}_{n\text{-times}} = q^n\, \mathbf{p}\, \left(\overline{q}\right)^n \qquad [2.6]$$

As we would suspect, this is a rotation of $\mathbf{p}$ around the axis $\boldsymbol{\mu}_q$ with an angle $n\phi_q/2$. The fact that the axis is kept the same during the successive rotation operations and that only the angle is affected is a consequence of the de Moivre formula given in [1.71].

Finally, it must be noted that, as is well known in Euclidean geometry, rotations can be expressed using reflections. Using the previously introduced reflection operator, we can see that two successive reflections with different axis/plane represent a rotation in the following way:

$$\mathcal{P}_{\boldsymbol{\eta}_2}\left[\mathcal{P}_{\boldsymbol{\eta}_1}\left[\mathbf{p}\right]\right] = \boldsymbol{\eta}_2\left(\boldsymbol{\eta}_1\mathbf{p}\boldsymbol{\eta}_1\right)\boldsymbol{\eta}_2 = \boldsymbol{\eta}_2\boldsymbol{\eta}_1\,\mathbf{p}\,\overline{\left(\boldsymbol{\eta}_2\boldsymbol{\eta}_1\right)} = \mathcal{R}_{\boldsymbol{\eta}_2\boldsymbol{\eta}_1}[\mathbf{p}] \qquad [2.7]$$

where the unit quaternion that encodes the rotation is $q = \boldsymbol{\eta}_2\boldsymbol{\eta}_1$. We made use of the property shared by pure quaternion: $\overline{\boldsymbol{\eta}} = -\boldsymbol{\eta}$ together with the anti-involution property of the quaternion conjugation. Note that as $\boldsymbol{\eta}_1$ and $\boldsymbol{\eta}_2$ are pure unit quaternions, then $q = \boldsymbol{\eta}_2\boldsymbol{\eta}_1$ is a *unit full quaternion*, which is necessary to have a 3D rotation. It is also necessary to have $\boldsymbol{\eta}_1 \neq \boldsymbol{\eta}_2$ in order to avoid degenerate cases.

Also note that if $\boldsymbol{\eta}_1 \perp \boldsymbol{\eta}_2$, then their product is a pure quaternion, i.e. $\boldsymbol{\eta}_1\boldsymbol{\eta}_2 = \boldsymbol{\eta}_3$ such that $\{\boldsymbol{\eta}_1, \boldsymbol{\eta}_2, \boldsymbol{\eta}_3\}$ is a triad. The rotation obtained by succesive reflections with axis $\boldsymbol{\eta}_1$ and $\boldsymbol{\eta}_2$ is then a special case of the rotation operator obtained in equation [2.7], where the angle of rotation is equal to π. Using the rotation operator notation from equation [2.3], it means that $q = \boldsymbol{\eta}_3$ and the angle of q is $\phi_q = \pi/2$. This involves that $\cos\phi_q = 0$ and thus that $q = \boldsymbol{\eta}_3$ is a pure unit quaternion. The rotation operator consists of a rotation around $\boldsymbol{\eta}_3$ with an angle of π.

2.1.3. *3D shears*

As introduced in [ELL 07a], we present here the quaternion formalism for the shear operator. In particular, two types of shears are studied here: axial-shears and beam-shears. The simplest shears are the axial-shears. They are defined for a 3D vector p (again considered as a pure quaternion) using two parameters, a scalar α (the shear factor) and a pure unit quaternion $\boldsymbol{\mu}_3$ (the shear axis), and take the form:

$$\mathcal{S}_{\alpha,\boldsymbol{\mu}_3}[\mathbf{p}] = \mathbf{p}\left(1 - \frac{\alpha}{2}\boldsymbol{\mu}_3\right) - \frac{\alpha}{2}\boldsymbol{\mu}_2\,\mathbf{p}\boldsymbol{\mu}_1 \qquad [2.8]$$

with $\boldsymbol{\mu}_1 \perp \boldsymbol{\mu}_2 \perp \boldsymbol{\mu}_3$ and $\boldsymbol{\mu}_1\boldsymbol{\mu}_2 = \boldsymbol{\mu}_3$, so that $\{1, \boldsymbol{\mu}_1, \boldsymbol{\mu}_2, \boldsymbol{\mu}_3\}$ form a quaternion basis. This means that the three pure unit quaternions form a right-handed triad. The shear operator reduces to the identity operator when $\alpha = 0$ and reversing α inverts the shear operation.

Now, the next level for shears is the beam-shear, that was introduced in [CHE 00]. This shear operator is parameterized by two scalar factors and a pure unit quaternion, part of a triad. The beam-shear operator reads:

$$\mathcal{S}_{\alpha,\beta,\boldsymbol{\mu}_3}[\mathbf{p}] = \mathbf{p}\left(\frac{1}{2} + \frac{\alpha}{2}\boldsymbol{\mu}_3\right) - \frac{\alpha}{2}\boldsymbol{\mu}_2\,\mathbf{p}\,\boldsymbol{\mu}_1 + \left(\frac{1}{2} + \frac{\beta}{2}\boldsymbol{\mu}_2\right)\mathbf{p} - \frac{\beta}{2}\boldsymbol{\mu}_1\,\mathbf{p}\,\boldsymbol{\mu}_3 \qquad [2.9]$$

where again $\boldsymbol{\mu}_1, \boldsymbol{\mu}_2, \boldsymbol{\mu}_3$ form a right-handed triad.

2.1.4. *3D dilations*

The quaternion formalism for dilations was first introduced in [ELL 07a]. Two types of dilations are presented here: axial- and radial-dilations.

Considering a 3D vector $\mathbf{p}$, the axial-dilation operator, denoted by $\mathcal{D}^a$, takes the form:

$$\mathcal{D}^a_{\boldsymbol{\mu},\alpha}[\mathbf{p}] = \frac{1+\alpha}{2}\mathbf{p} + \frac{1-\alpha}{2}\boldsymbol{\mu}\,\mathbf{p}\,\boldsymbol{\mu} \qquad [2.10]$$

where $\boldsymbol{\mu}$ is the axis of dilation/compression. The factor α controls the behavior of the transformation, leading to a compression if $|\alpha| < 1$, a dilation if $|\alpha| > 1$ and identity operator for $\alpha = 1$. Negative values of α result in a reversal of the vector $\mathbf{p}$.

For the radial dilation case, the operator is denoted by $\mathcal{D}^r$ and takes the form:

$$\mathcal{D}^r_{\boldsymbol{\mu},\alpha}[\mathbf{p}] = \frac{2+\alpha}{2}\mathbf{p} + \frac{\alpha}{2}\boldsymbol{\mu}\,\mathbf{p}\,\boldsymbol{\mu} \qquad [2.11]$$

This operator expands space outward from an invariant line defined by $\boldsymbol{\mu}$. Here, dilations occur for $\alpha > 0$ and compression occurs for $\alpha < 0$, while $\alpha = 0$ defines the identity operator.

2.1.5. *4D reflections*

As mentioned previously, quaternions have the ability to encode 4D geometric transformations [COX 46]. Here, we introduce the 4D reflection operator for a 4D vector p, encoded as a full quaternion, i.e. $p \in \mathbb{H}$. The reflection of p across the hyperplane defined by its normal 4D vector q (a *unit full* quaternion, i.e. $q \in \mathbb{H}$ and $|q| = 1$) is given by:

$$\mathcal{P}_q[p] = -q\,\overline{p}\,q \qquad [2.12]$$

Just like in the 3D case in section 2.1.1, the 4D reflection operator is its own inverse since:

$$\mathcal{P}_q[\mathcal{P}_q[p]] = q\overline{(q\,\overline{p}\,q)}\,q = q\,\overline{q}\,p\,\overline{q}\,q = p \qquad [2.13]$$

where we used the fact that q is a unit quaternion, i.e. $\overline{q}q = q\overline{q} = 1$.

2.1.6. *4D rotations*

The last Euclidean transformation is the 4D rotation of a 4D vector p (encoded as a full quaternion, i.e. $p \in \mathbb{H}$). The 4D rotation operator is encoded using a pair of full unit quaternions q and r, i.e. $q, r \in \mathbb{H}$ and $|q| = |r| = 1$, such that $\mathrm{S}(q) = \mathrm{S}(r)$ [COX 46]. The rotation operator takes the form:

$$\mathcal{R}_{q,r}[p] = q\,p\,r \qquad [2.14]$$

where the rotation $\mathcal{R}_{q,r}[.]$ stands for a rotation about a plane (to be identified later) with angle 2ϕ if $\mathrm{S}(q) = \mathrm{S}(r) = \cos\phi$. Note that the scalar parts of q and r are cosines due to the fact that $|q| = |r| = 1$.

Now, just as in the 3D case, a rotation can be expressed as a combination of two reflections. In the 4D case, it arises in the following way when considering the product of two reflections across planes with h_1 and h_2, respectively, as normal vector (with the requirement that $h_1 \neq h_2$):

$$\mathcal{P}_{h_2}\left[\mathcal{P}_{h_1}[p]\right] = h_2 \, \overline{(h_1 \, \overline{p} \, h_1)} \, h_2 = h_2 \, \overline{h_1} \, p \, \overline{h_1} \, h_2 \qquad [2.15]$$

As h_1 and h_2 are pure unit quaternions, their product is a full unit quaternion. The quaternions $h_2 \, \overline{h_1}$ and $\overline{h_1} \, h_2$ are different (multiplication order); however, they have their scalar part in common, which is not affected by the order of quaternion multiplication, so that $\mathrm{S}(h_2 \, \overline{h_1}) = \mathrm{S}(\overline{h_1} \, h_2)$. This, in addition to the unit norm of the two quaternion products, is the requirement for the operator to be a 4D rotation. In fact, this rotation operator performs a rotation of angle ϕ with $\cos\phi = \mathrm{S}(h_2 \, \overline{h_1}) = \mathrm{S}(\overline{h_1} \, h_2)$ about a plane defined by $\mathrm{S}(h_2 \, \overline{h_1 p}) = \mathrm{S}(\overline{h_1} \, h_2 \overline{p}) = 0$. We can also identify the obtained quaternions with the elements of the 4D rotation operator in section 2.14 like: $h_2 \, \overline{h_1} = q$ and $\overline{h_1} \, h_2 = r$. The interested readers can refer to [COX 46, COX 74] for more on 4D Euclidean geometry.

2.2. Spherical geometry

Geometry on a sphere resembles geometry on a plane. Thus, if $\mathcal{A}$ and $\mathcal{B}$ are two points on a sphere, then in analogy to a segment on a plane one can consider $\widehat{\mathcal{AB}} = \mathcal{B} - \mathcal{A}$ to be the distance between the points $\mathcal{A}$ and $\mathcal{B}$, or the arc of a great circle running from point $\mathcal{A}$ to point $\mathcal{B}$. The idea of the difference of two points as an arc joining them is completely analogous to the difference between two points in a plane being the vector joining them. Hence, two arcs will be equal if and only if they lie on the same great circle, and their lengths and directions are equal. Two equal arcs can be superimposed by moving one along the great circle on which they are located. One difference between planar and spherical geometry is that the *addition* of arcs is in general non-commutative. The addition of arcs is commutative only if they lie on the same great circle, in which case the addition reduces to a simple arithmetic sum.

Branets [BRA 74], following Hamilton, considered the unit quaternion λ that can be represented in the form of a quotient of two equal-length pure quaternions or vectors in 3D-space, a and b, i.e.,

$$\lambda = b \, a^{-1} = cos\phi + \boldsymbol{\mu} sin\phi. \qquad [2.16]$$

The reciprocal of a is defined through [1.26]. It can be shown that the following conditions hold:

1) the angle ϕ from the vector a to b satisfies $\phi \leq \pi$;

2) the unit vector $\boldsymbol{\mu}$ is normal to the plane containing a and b;

3) the vectors a, b and $\boldsymbol{\mu}$ span all of $\mathbb{R}^3$.

In other words, if we associate the point $\mathcal{A}$ on a sphere with the vector a, and the point $\mathcal{B}$ with the vector b, then the angle of the arc $\widehat{\mathcal{AB}}$ is in fact ϕ given in [2.16]. This situation is depicted in Figure 2.1. This implies that the quaternion product of two normalized quaternions corresponds to the geometric sum of the arcs of a sphere of radius $|a| = |b|$. So we have at our disposal the equivalence between arcs and unit quaternions, denoted as $\widehat{\mathcal{AB}} \iff \lambda_{b,a}$, which allows the correspondence

$$\widehat{\mathcal{AB}} + \widehat{\mathcal{BC}} = \widehat{\mathcal{AC}} \iff \lambda_{c,b}\,\lambda_{b,a} = \lambda_{c,a}$$

where we have included a third point $\mathcal{C}$ with its vector c.

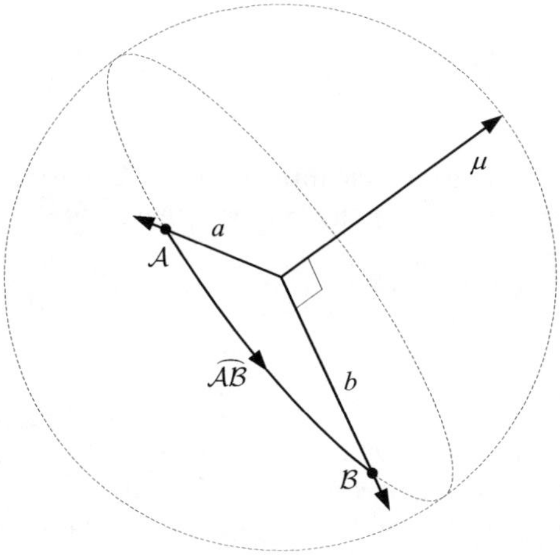

Figure 2.1. *Spherical geometry as unit length quaternion*

Conversely, any unit quaternion λ defines an arc of a great circle, whose plane is normal to $\boldsymbol{\mu}$ and whose arc length is ϕ; the position of the arc on its great circle is arbitrary, i.e. the arc may be slid along the circle.

Conjugating the quaternion $\overline{\lambda}$ reverses the direction of the arc along the same great circle. Negating the quaternion $-\lambda$ corresponds to the arc from $-a$ to b.

The special cases of an arc of length zero corresponds to a quaternion $\lambda = 1 + 0\boldsymbol{\mu}$ so that the plane of the arc is arbitrary. The case where the arc length is π corresponds to $\lambda = -1 + 0\boldsymbol{\mu}$, as again the direction of the arc is arbitrary when moving a point to its diametrically opposite point across the sphere – any direction will do. These associations provide a mechanism for encoding spherical triangles. The geometric sum of the three arcs $\widehat{\mathcal{AB}}$, $\widehat{\mathcal{BC}}$ and $\widehat{\mathcal{CA}}$ forming a closed spherical triangle, when mapped to their unit quaternions, is:

$$\lambda_{a,c}\,\lambda_{c,b}\,\lambda_{b,a} = (a\,c^{-1})(c\,b^{-1})(b\,a^{-1}) = a\,(c^{-1}c)(b^{-1}b)\,a^{-1} = 1$$

which is the arc of zero length. In general, unit quaternions which represent arcs that are cyclically arranged edges of a closed polygon have a quaternion product equal to 1.

2.3. Projective space (3D)

Full quaternions, those with non-zero scalar part, can be used to represent *points* in homogeneous coordinates. The quaternion identity:

$$q = s + \mathbf{v} = s\left[1 + \frac{\mathbf{v}}{s}\right] \qquad [2.17]$$

where $s = \mathrm{S}(q)$ and $\mathbf{v} = \mathbf{V}(q)$. In this form, q can be used to represent a point located at the end of the vector $p = \mathbf{v}/s$ from the origin with weight s. Thus, in weighted-point form:

$$q = s[1 + p]$$

Hence, the set of quaternions can be associated with the real projective space $\mathbb{P}^3 \cong \mathbb{R}^4$. This interpretation was discussed by Joly [JOL 05] and MacFarlane [MAC 06] in 1905 and 1906, respectively. Under this interpretation, a unit-weight point at the origin is denoted as $q = 1[1 + 0\boldsymbol{i} + 0\boldsymbol{j} + 0\boldsymbol{k}]$, and a *weightless* point at infinity in the (x, y, z) direction is denoted as $q = [0 + x\boldsymbol{i} + y\boldsymbol{j} + z\boldsymbol{k}]$, which can be seen by letting $s \to 0$ in [2.17]. The resulting pure quaternion can also be viewed as a translation *vector* when added to a weighted-point.

The algebra of the set of vectors $\mathcal{V} \cong \mathbf{V}(\mathbb{H})$ and the algebra of the set of (weighted) points $\mathcal{P} \cong \mathbb{H}$, both described using quaternion algebra, are different in subtle ways. For example, the direct sum of two vectors, $\mathbf{v}_1$ and $\mathbf{v}_2$, follows the *parallelogram law*, whereas the direct sum of two (weighted) points, p_1 and p_2, follows the *principle of center-of-mass*. This can be seen in the following equation:

$$q_1 + q_2 = \underbrace{(s_1 + s_2)}_{\text{weight}}[1 + \underbrace{(s_1 p_1 + s_2 p_2)\,/\,(s_1 + s_2)}_{\text{center-of-mass}}]$$

Likewise, scaling a vector $\mathbf{v} \in \mathcal{V}$ by $\alpha \in \mathbb{R}$ changes its length but leaves its direction unchanged, whereas scaling a point $p \in \mathcal{P}$ leaves its position unchanged but changes its weight since $\alpha q = \alpha s[1 + p]$.

The set of vectors is closed with respect to quaternion addition and subtraction; however, the set of weighted-points is not closed. For example, the difference between two equal-weight points, q_1 and q_2, is the *weightless* point (i.e. the vector) that joins them as:

$$q_1 - q_2 = s\,[1 + p_1] - s\,[1 + p_2] = 0 + (p_1 - p_2)$$

Table 2.1 provides a summary of how various quaternion algebraic operations map elements between point and vector subsets.

Mapping	Projective space interpretation
$\alpha\mathcal{V} \to \mathcal{V}$	Scales vector length, direction unchanged.
$\alpha\mathcal{P} \to \mathcal{P}$	Scales point weight, position unchanged.
$\mathcal{V} \pm \mathcal{V} \to \mathcal{V}$	Parallelogram law.
$\mathcal{P} \pm \mathcal{V} \to \mathcal{P}$	Translates point in direction of vector.
$\mathcal{P} + \mathcal{P} \to \mathcal{P}$	Principle of center-of-mass.
$\mathcal{P} - \mathcal{P} \to \mathcal{V}$ or $\mathcal{P}$	$\mathcal{V}$ iff weights are equal.
$\mathcal{V}\mathcal{V} \to \mathcal{V}$ or $\mathcal{P}$	$\mathcal{V}$ iff vectors are perpendicular.
$\mathcal{P}\mathcal{P} \to \mathcal{V}$ or $\mathcal{P}$	$\mathcal{V}$ iff $\langle p_1, p_2 \rangle = 1$.

Table 2.1. *Summary of point ($\mathcal{P} \cong \mathbb{H}$) and vector ($\mathcal{V} \cong \mathbf{V}(\mathbb{H})$) mappings with $\alpha \in \mathbb{R}$*

Quaternion conjugation, by definition, negates the vector part of a quaternion. Consequently, conjugation and negation of a vector are the same, i.e. $\bar{\mathbf{v}} = -\mathbf{v}$. In contrast, the conjugate of a weighted-point results in the point's position being reflected across the origin, whereas negation changes the sign of the point's weight, but its position remains the same, i.e.: $\overline{s\,[1 + p]} = s\,[1 - p] \neq -s\,[1 + p_2]$.

Not withstanding these differences, some algebraic combinations behave similarly for vectors and points. For example, given two vectors, $\mathbf{v}_1$ and $\mathbf{v}_2$, their linearly weighted sum:

$$v_\alpha = (1 - \alpha)\,\mathbf{v}_1 + (\alpha)\,\mathbf{v}_2$$

traces a line from $\mathbf{v}_1$ to $\mathbf{v}_2$ as α varies from 0 to 1, i.e. it draws a line diagonally across the parallelogram defined by $\mathbf{v}_1$ and $\mathbf{v}_2$. Likewise, given two points, p_1 and p_2, their linearly weighted sum:

$$q_\alpha = (1 - \alpha)\,s_1\,[1 + p_1] + (\alpha)\,s_2\,[1 + p_2] = s_1'\,[1 + p_1] + s_2'\,[1 + p_2]$$

also traces the same path across the diagonal of the parallelogram defined by points p_1 and p_2. However, for a constant rate of change in α the weighed points are not in general equally spaced, as it is in the vector case, due to the relative weight of the two points. The points would be equally spaced only if their weights are equal. See Figure 2.2 for an example comparison where p_2 has a weight three times that of p_1.

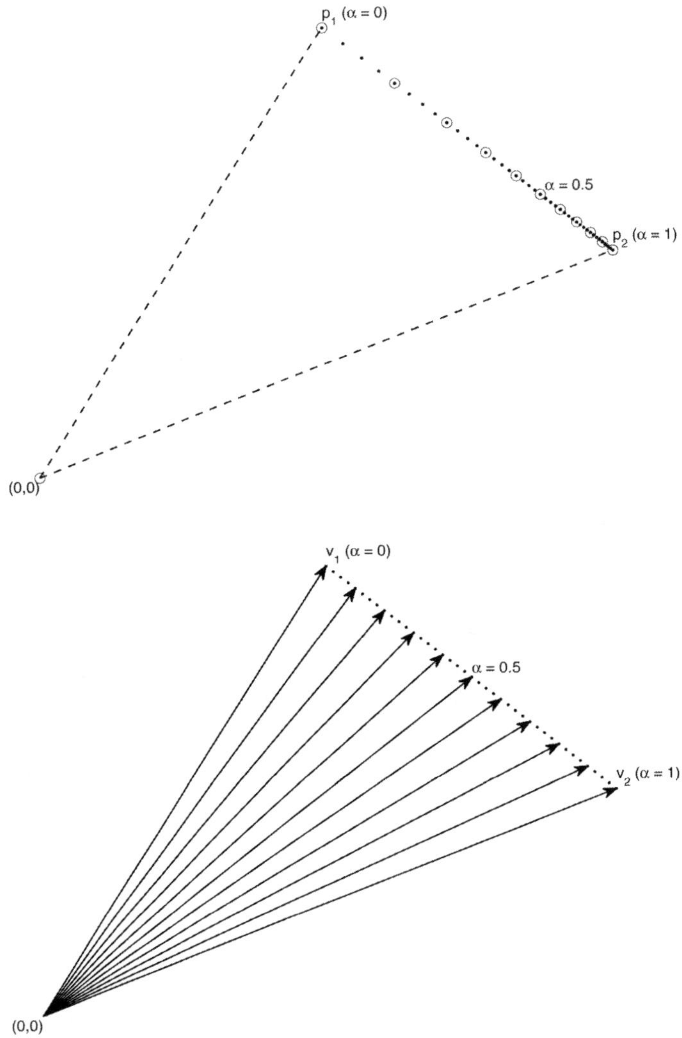

Figure 2.2. *Linear weighted sum of points (top) versus vectors (bottom)*

Even though points and vectors are disjoint subsets, they share a common framework when viewed as a projective space. A vector represents a point at infinity along its direction so that geometrically all the points at infinity have been included into a single 3-space model. We may ask, "what advantage does this projective space model of separating points and vectors grant us?" We saw in section 2.1 the use of various linear quaternion functions that map vectors into vectors by way of rotations, reflections, shears and dilations. This set of *linear* operations can now be expanded into projections (including translations). To understand this, we must digress and discuss the most general form of linear quaternion equation.

2.3.1. *Systems of linear quaternion functions*

Real linear functions take the monomial form: $f(x) = mx$, where $m, x \in \mathbb{R}$. Linear combinations, i.e. direct sums or compositions, of such functions can always be reduced to this same form. In contrast, quaternion linear functions have the multinomial form:

$$f(q) = \sum_{p=1}^{P} m_p \, q \, n_p \qquad [2.18]$$

where all factors are quaternion-valued, i.e., $q, m_p, n_p \in \mathbb{H}$. We would expect under functional compositions that the number of terms in the summation can become arbitrarily large due to quaternion multiplication being non-commutative. However, this general multinomial can always be reduced to its *quaternary canonical form*:

$$f(q) = A\,q + B\,q\,\boldsymbol{i} + C\,q\,\boldsymbol{j} + D\,q\,\boldsymbol{k}, \qquad [2.19]$$

where $A, B, C, D \in \mathbb{H}$.

The linear sum of two such functions is given by the component-wise addition. Let $f_1(q) = A_1 q + B_1 q\,\boldsymbol{i} + C_1 q\,\boldsymbol{j} + D_1 q\,\boldsymbol{k}$ and $f_2(q) = A_2 q + B_2 q\,\boldsymbol{i} + C_2 q\,\boldsymbol{j} + D_2 q\,\boldsymbol{k}$ then:

$$f_1(q) + f_2(q) = A_3 q + B_3 q\,\boldsymbol{i} + C_3 q\,\boldsymbol{j} + D_3 q\,\boldsymbol{k}$$

where $A_3 = A_1 + A_2$, $B_3 = B_1 + B_2$, $C_3 = C_1 + C_2$ and $D_3 = D_1 + D_2$.

The composition of two linear functions, $f_2(f_1(q))$, is given by:

$$f_2(f_1(q)) = A_3 q + B_3 q\boldsymbol{i} + C_3 q\boldsymbol{j} + D_3 q\boldsymbol{k}$$

where:

$$\begin{aligned}
A_3 &= A_2A_1 - B_2B_1 - C_2C_1 - D_2D_1, \\
B_3 &= A_2B_1 + B_2A_1 - C_2D_1 + D_2C_1, \\
C_3 &= A_2C_1 + B_2D_1 + C_2A_1 - D_2B_1, \\
D_3 &= A_2D_1 - B_2C_1 + C_2B_1 + D_2A_1.
\end{aligned}$$

Careful examination of this composition rule reveals that it has the same structure as the standard quaternion multiplication. The composition is, of course, not commutative: $f_2(f_1(q)) \neq f_1(f_2(q))$. This is because geometrical operations do not commute (the composition of two rotations, for example, gives different results, in general, according to the order in which the rotations are carried out).

The equivalence between invertible functions in quaternary canonic form and the general linear group of 4×4 invertible real matrices, $GL_4(\mathbb{R})$, will be used later. Let $p = p_0 + p_1\boldsymbol{i} + p_2\boldsymbol{j} + p_3\boldsymbol{k}$ and $q = q_0 + q_1\boldsymbol{i} + q_2\boldsymbol{j} + q_3\boldsymbol{k}$ be two arbitrary quaternions. Using standard hypercomplex operator product rules, the components of the quaternion product $pq = q'_0 + q'_1\boldsymbol{i} + q'_2\boldsymbol{j} + q'_3\boldsymbol{k} = q'$ are:

$$\begin{aligned}
q'_0 &= p_0q_0 - p_1q_1 - p_2q_2 - p_3q_3, \\
q'_1 &= p_1q_0 + p_0q_1 - p_3q_2 + p_2q_3, \\
q'_2 &= p_2q_0 + p_0q_2 + p_3q_1 - p_1q_3, \\
q'_3 &= p_3q_0 + p_0q_3 - p_2q_1 + p_1q_2.
\end{aligned}$$

By gathering terms into matrix-vector notation:

$$\begin{bmatrix} q'_0 \\ q'_1 \\ q'_2 \\ q'_3 \end{bmatrix} = \begin{bmatrix} p_0 & -p_1 & -p_2 & -p_3 \\ p_1 & p_0 & -p_3 & p_2 \\ p_2 & p_3 & p_0 & -p_1 \\ p_3 & -p_2 & p_1 & p_0 \end{bmatrix} \begin{bmatrix} q_0 \\ q_1 \\ q_2 \\ q_3 \end{bmatrix} \qquad [2.20]$$

we can define an equivalent matrix-vector form for the quaternion product. Let $[\![p]\!]$ and $[q]$ denote the matrix and vector equivalences, respectively. Since the quaternion product is bilinear, we can instead place the components of p into the vector and the components of q into the matrix, but in doing so the lower right 3×3 submatrix is transposed from the one given above. Denote this *transmuted* form of the matrix as $[\![q]\!]^{\dagger}$. Therefore, any quaternion product has two equivalent matrix-vector forms:

$$pq \leftrightarrow [\![p]\!]\,[q] = [\![q]\!]^{\dagger}\,[p]\,.$$

Hence, any triple-product pqr may be arbitrarily reordered in matrix form as

$$pqr \leftrightarrow [\![p]\!]\,[\![q]\!]\,[r] = [\![p]\!]\,[\![r]\!]^{\dagger}\,[q]$$

The linear quaternary equation $q' = Aq + Bq\boldsymbol{i} + Cq\boldsymbol{j} + Dq\boldsymbol{k}$ can be written in matrix-vector form as

$$[q'] = \left\{[\![A]\!] + [\![B]\!]\,[\![i]\!]^{\dagger} + [\![C]\!]\,[\![j]\!]^{\dagger} + [\![D]\!]\,[\![k]\!]^{\dagger}\right\}[q] = [\![Q]\!]\,[q] \qquad [2.21]$$

Now, both a matrix $[\![p]\!]$ and a transmuted matrix $[\![q]\!]^{\dagger}$ contain only four degrees of freedom: one for each component of the quaternion. Careful examination of the four terms in the above matrix equation reveals that these matrices are independent, giving the total of 16 degrees of freedom necessary for an arbitrary matrix. It is straightforward to start with an arbitrary matrix $[\![Q]\!]$ and solve for the elements of the four quaternions $\{A, B, C, D\}$. For details of this analysis, see [ELL 07b].

2.3.2. *Projective transformations*

Having established the connection between $GL_4(\mathbb{R})$, the general linear group of 4×4 invertible real matrices, we can now discuss the various types of geometrical transformation possible using a linear quaternion function. We draw here on the classic text of Bruce Meserve [MES 83] which clearly sets out the possibilities (although not using a quaternion formalism as given here).

The most general transformation is *projective*. The *affine* transformations are a subset of the *projective* transformations with the property that straight lines are preserved (that is, remain straight) under the transformation. Reflection, scaling and shear are examples of affine transformations. *Similarity* transformations are a subset of the affine transformations with the additional property that angles are preserved (thus geometric figures are transformed into similar geometric figures). Uniform scaling (that is, scaling along all axes by the same scale factor) has this property. *Euclidean* transformations are a subset of the similarity transformations. Translation and rotation are examples. They preserve angles as well as straight lines.

Two additional subsets of the affine transformations "cut across" the other subsets. The *isometries* preserve the distance between points. The *linear* transformations are those that obey superposition. In geometric terms, it means we may decompose a vector or point into the sum of two or more vectors or points and obtain the same results by transforming the decomposed parts separately as by transforming the undecomposed vectors or points.

Now a very significant point is that all these transformations are linear when expressed as operations on a projective space, even if non-linear in the 3D Euclidean space of the signal samples or image pixels. All of these linear transforms can be encoded using real 4×4 matrices as shown in Figure 2.3, which were shown to be equivalent to linear quaternion functions in section 2.3.1.

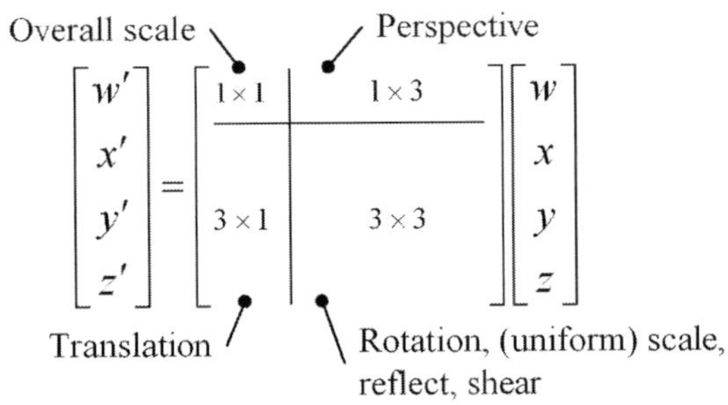

Figure 2.3. *Projective maps using real* 4×4 *matrix*

3

Quaternion Fourier Transforms

Quaternion Fourier transforms provide elegance and expressive power in the analysis of vector-valued signals and images. There is, however, a cost – an overwhelming number of transform definitions. This chapter provides some of the possible quaternion Fourier transforms definitions and their properties which appear best suited to vector-valued signals.

The classical (complex) Fourier transform of a one-dimensional (1D) complex-valued function $f(t) : \mathbb{R} \to \mathbb{C}$ can be defined as:

$$\mathcal{F}\{f\}(\omega) = F(\omega) = \int_{-\infty}^{\infty} f(t)\, e^{-I\omega t}\, \mathrm{d}t, \qquad [3.1]$$

where I is the complex square root of -1. The inverse transform is defined by conjugating the exponential kernel and adding a scale factor as in

$$\mathcal{F}^{-1}\{F\}(\omega) = f(t) = \frac{1}{2\pi}\int_{-\infty}^{\infty} F(\omega)\, e^{+I\omega t}\, \mathrm{d}\omega. \qquad [3.2]$$

Symbolically, we write the signal and its transform pair as $f(t) \rightleftharpoons F(\omega)$. The extension of the complex Fourier transform to two dimensions is straightforward. Given a signal $f(x, y)$: $\mathbb{R}^2 \to \mathbb{C}$, the two-dimensional (2D) complex Fourier transform is

$$\mathcal{F}\{f\}(\omega, \boldsymbol{\nu}) = F(\omega, \boldsymbol{\nu}) = \int_{-\infty}^{\infty}\int_{-\infty}^{\infty} f(x, y) e^{-I(\omega x + \boldsymbol{\nu} y)}\, \mathrm{d}x\, \mathrm{d}y$$

The index space of the transformed function is of course 2D, like the original function. As in the 1D case, the inverse transform is defined by conjugating the exponential kernel and adding a scale factor as in

$$\mathcal{F}^{-1}\{F\}(x,y) = f(x,y) = \frac{1}{(2\pi)^2}\int_{-\infty}^{\infty}\int_{-\infty}^{\infty} F(\omega,\boldsymbol{\nu})e^{+I(\omega x+\boldsymbol{\nu} y)}\,\mathrm{d}\omega\,\mathrm{d}\boldsymbol{\nu}.$$

A sufficient (though not necessary) condition that a function $f(t)$ has a complex Fourier transform is that the function should be absolutely integrable, i.e.,

$$\int_{\mathbb{R}} |f(t)|\,\mathrm{d}t < \infty \qquad [3.3]$$

The set of such functions is denoted by $L^1(\mathbb{R};\mathbb{C})$. To be able to recover the original function from the inverse Fourier transform, the function must be of bounded variation, which means that $f(t)$ can be expressed as the difference of two bounded monotonically increasing functions. Alternately, if the function is square integrable, i.e.,

$$\int_{\mathbb{R}} |f(t)|^2\,\mathrm{d}t < \infty, \qquad [3.4]$$

then both the Fourier transform and its inverse exist. Square integrable functions are denoted by $L^2(\mathbb{R};\mathbb{C})$.

To generalize the Fourier transform to quaternion-valued signals, the approach taken in this work is to replace the *single* complex square root of $-1\ (= I^2)$ with one of many possible quaternion square roots of $-1 = \boldsymbol{\mu}^2$. In the quaternion domain, $\boldsymbol{\mu}$ is taken from the set of all pure unit length quaternions, namely $\boldsymbol{\mu} \in \{q \in \mathbf{V}(\mathbb{H}) : \|q\| = 1\}$. This provides an additional degree of design freedom, beyond the complex Fourier transform, when treating problems in vector signal processing and can be chosen to advantage for each application. The ability to choose the root of -1 is particularly useful when dealing with 2D signals where each integral may be associated with its own root, or not. For purposes of later comparison, Table 3.1 contains the basic properties of the complex Fourier transform used in image and signal processing.

Table 3.1 contains two operator formulas, which are the building blocks of image and signal processing: convolution and correlation. The formal definition of a convolution of two functions is

$$(f_1 * f_2)(t) = \int_{-\infty}^{+\infty} f_1(\tau)f_2(t-\tau)\,\mathrm{d}\tau.$$

$\frac{1}{\sqrt{2\pi}}\int_{-\infty}^{+\infty} e^{I\omega t}F(\omega)\,\mathrm{d}\omega = f(t)$		$\rightleftharpoons$	$F(\omega) = \frac{1}{\sqrt{2\pi}}\int_{-\infty}^{+\infty} e^{-I\omega t}f(t)\,\mathrm{d}t$	
Time-domain	$f(t)$	$\rightleftharpoons$	$F(\omega)$	Frequency-domain
Time reversal	$f(-t)$	$\rightleftharpoons$	$F(-\omega)$	Frequency reversal
Complex conjugation	$f^\star(t)$	$\rightleftharpoons$	$F^\star(-\omega)$	Reversed conjugation
Reversed conjugation	$f^\star(-t)$	$\rightleftharpoons$	$F^\star(\omega)$	Complex conjugation
Purely real	$f(t) \in \mathbb{R}$	$\rightleftharpoons$	$F(\omega) = F^\star(-\omega)$	Even, symmetric
Purely imaginary	$f(t) \in I\mathbb{R}$	$\rightleftharpoons$	$F(\omega) = -F^\star(-\omega)$	Odd, anti-symmetric
Even, symmetric	$f(t) = f^\star(-t)$	$\rightleftharpoons$	$F(\omega) \in \mathbb{R}$	Purely real
Odd, anti-symmetric	$f(t) = -f^\star(-t)$	$\rightleftharpoons$	$F(\omega) \in I\mathbb{R}$	Purely imaginary
Time shift	$f(t-\tau)$	$\rightleftharpoons$	$F(\omega)\,e^{-I\omega\tau}$	Modulation
Modulation	$f(t)\,e^{I\boldsymbol{\nu}t}$	$\rightleftharpoons$	$F(\omega-\boldsymbol{\nu})$	Frequency shift
Time scale	$f(\alpha t)$	$\rightleftharpoons$	$\frac{1}{\lvert\alpha\rvert}F(\frac{\omega}{\alpha})$	
	$\frac{1}{\lvert\alpha\rvert}f(\frac{t}{\alpha})$	$\rightleftharpoons$	$F(\alpha\omega)$	Frequency scale
Linearity	$\alpha f_1(t) + \beta f_2(t)$	$\rightleftharpoons$	$\alpha F_1(\omega) + \beta F_2(\omega)$	
Multiplication	$f_1(t)\,f_2(t)\,\sqrt{2\pi}$	$\rightleftharpoons$	$F_1(\omega) * F_2(\omega)$	Convolution
Convolution	$(f_1 * f_2)(t)$	$\rightleftharpoons$	$\sqrt{2\pi}\,F_1(\omega)\,F_2(\omega)$	Multiplication
Correlation	$(f_1 \star f_2)(t)$	$\rightleftharpoons$	$\sqrt{2\pi}\,F_1^*(\omega)F_2(\omega)$	
Delta function	$\delta(t)\,\sqrt{2\pi}$	$\rightleftharpoons$	1	
Shifted delta	$\delta(t-\tau)\,\sqrt{2\pi}$	$\rightleftharpoons$	$e^{-I\omega\tau}$	
	$e^{I\boldsymbol{\nu}t}$	$\rightleftharpoons$	$\sqrt{2\pi}\,\delta(\omega-\boldsymbol{\nu})$	Shifted delta
	1	$\rightleftharpoons$	$\sqrt{2\pi}\,\delta(\omega)$	Delta function
n*th* derivative	$\frac{\mathrm{d}^n}{\mathrm{d}t^n}f(t)$	$\rightleftharpoons$	$(I\omega)^n F(\omega)$	
	$t^n f(t)$	$\rightleftharpoons$	$I^n \frac{\mathrm{d}^n}{\mathrm{d}\omega^n}F(\omega)$	n*th* derivative

NOTE.– $f(t) : \mathbb{R} \to \mathbb{C}, \quad \alpha, \beta \in \mathbb{R}.$

Table 3.1. *1D Complex Fourier transform pairs*

Likewise, the cross-correlation of two functions is defined as:

$$(f_1 \star f_2)(t) = \int_{-\infty}^{+\infty} f_1^\star(\tau) f_2(t+\tau)\,\mathrm{d}\tau.$$

The correlation is superficially similar to convolution, but it has a different physical interpretation. The generalization from complex to quaternion algebra comes at a cost: neither the correlation nor the convolution are commutative, and which function is shifted (f_1 or f_2) changes the results. In the complex domain, these differences are immaterial, but with quaternions it means that the formulas given in Table 3.1 do not work: instead we must find the correct quaternion generalizations, as discussed in section 3.1.5.

3.1. 1D quaternion Fourier transforms

In the complex case, the sign of the exponents of [3.1] and [3.2] must differ. Which of the two you choose to be positive does not matter, so long as you keep to the rule. If someone else has used the opposite choice, the Fourier pair calculated from a given function will be the complex conjugate of that given by your choice. This simple rule does not extend to the quaternion case. Still further definitions for the quaternion Fourier transform (QFT) arise from the fact that quaternion multiplication is non-commutative; hence, the placement of the exponential kernel in the transform definition becomes meaningful. This situation gives rise to four unique 1D QFTs.

3.1.1. *Definitions*

The 1D *right-sided* [1] QFTs of a quaternion-valued signal $f(t) : \mathbb{R} \to \mathbb{H}$ are:

$$\mathcal{F}^{R}_{\mp\boldsymbol{\mu}}\{f\}(\omega) = F^{R}(\omega) = \kappa_{-} \int_{-\infty}^{\infty} f(t)\, e^{\mp\boldsymbol{\mu}\omega t}\, \mathrm{d}t \qquad [3.5]$$

where the $-\boldsymbol{\mu}$ version is referred to as the *forward* transform and the $+\boldsymbol{\mu}$ version is the *reverse* transform, which is distinct from their *inverse* transforms given, respectively, by

$$\mathcal{F}^{-R}_{\mp\boldsymbol{\mu}}\{F^{R}\}(t) = f(t) = \kappa_{+} \int_{-\infty}^{\infty} F(\omega)\, e^{\pm\boldsymbol{\mu}\omega t}\, \mathrm{d}\omega. \qquad [3.6]$$

In both equations, $\boldsymbol{\mu} \in \mathbb{H}$ is a pure unit quaternion so that $\boldsymbol{\mu}^2 = -1$. The scale factors must satisfy: $\kappa_{+}\kappa_{-} = (2\pi)^{-1}$. If $\kappa_{-} = \kappa_{+}$, then the transform is a *unitary transform* with the consequence that the transform preserves the energy of the original signal $f(t)$ when it is mapped to the frequency domain[2].

Likewise, the *left-sided* QFTs of a quaternion-valued signal, $f(t) : \mathbb{R} \to \mathbb{H}$ are:

$$\mathcal{F}^{L}_{\mp\boldsymbol{\mu}}\{f\}(\omega) = F^{L}(\omega) = \kappa_{-} \int_{-\infty}^{\infty} e^{\mp\boldsymbol{\mu}\omega t}\, f(t)\, \mathrm{d}t, \qquad [3.7]$$

1 In this book, we use the term *right-sided* to refer to the placement of the exponential relative to the signal (the exponential is on the *right* for a right-sided transform). This is consistent with usage in the QTFM library [SAN 13b], where the transform handedness is selected by a parameter with values "L" or "R". Care is needed to check whether this convention is used when reading other quaternion literature, since the convention may differ.

2 The scale factor requirement also holds for complex Fourier transforms.

and their inverses are given by:

$$\mathcal{F}_{\mp\boldsymbol{\mu}}^{-L}\{F^L\}(\omega) = f(t) = \kappa_+ \int_{-\infty}^{\infty} e^{\pm\boldsymbol{\mu}\omega t}\, F^L(\omega)\, \mathrm{d}\omega. \qquad [3.8]$$

When the context makes it clear, or when either a left- or a right-sided transform may be used, the superscript L/R will be dropped.

Not all quaternion-valued functions can be Fourier-transformed. They are transformable if they fulfill certain conditions, known as the Dirichlet conditions. We need not use first principles to discover which functions are transformable. Instead, we can borrow from the classic complex case as follows. Split $f(t)$ via symplectic decomposition (equation [1.55]) into simplex and perplex parts with respect to $\boldsymbol{\mu}$, the axis of the transform, as:

$$f(t) = f_s(t) + f_p(t)\, \boldsymbol{\mu}_2, \quad \text{where } \boldsymbol{\mu}_2 \perp \boldsymbol{\mu} \text{ and } f_s, f_p \in \mathbb{C}_{\boldsymbol{\mu}}.$$

The left-sided QFT of $f(t)$ becomes:

$$F^L(\omega) = \kappa_- \int_{\mathbb{R}} e^{-\boldsymbol{\mu}\omega t} f_s(t)\, \mathrm{d}t + \kappa_- \int_{\mathbb{R}} e^{-\boldsymbol{\mu}\omega t} f_p(t)\, \mathrm{d}t \boldsymbol{\mu}_2. \qquad [3.9]$$

so that:

$$F^L(\omega) = F_s(\omega) + F_p(\omega)\boldsymbol{\mu}_2, \qquad [3.10]$$

where:

$$F_s(\omega) = \kappa_- \int_{\mathbb{R}} e^{-\boldsymbol{\mu}\omega t} f_s(t)\, \mathrm{d}t, \quad \text{and} \quad F_p(\omega) = \kappa_- \int_{\mathbb{R}} e^{-\boldsymbol{\mu}\omega t} f_p(t)\, \mathrm{d}t.$$

Likewise, the right-sided QFT of $f(t)$ becomes:

$$\begin{aligned} F^R(\omega) &= \kappa_- \int_{\mathbb{R}} f_s(t) e^{-\boldsymbol{\mu}\omega t}\, \mathrm{d}t + \kappa_- \int_{\mathbb{R}} f_p(t)\boldsymbol{\mu}_2 e^{-\boldsymbol{\mu}\omega t}\, \mathrm{d}t \qquad [3.11] \\ &= \kappa_- \int_{\mathbb{R}} f_s(t) e^{-\boldsymbol{\mu}\omega t}\, \mathrm{d}t + \kappa_- \int_{\mathbb{R}} f_p(t) e^{+\boldsymbol{\mu}\omega t}\, \mathrm{d}t\, \boldsymbol{\mu}_2 \end{aligned}$$

so that:

$$F^R(\omega) = F_s(\omega) + F_p(-\omega)\boldsymbol{\mu}_2. \qquad [3.12]$$

Due to the equivalence between $\mathbb{C}_{\boldsymbol{\mu}}$ and the standard complex numbers, and the fact that the QFT can be decomposed into a sum of complex subfield functions, the existence and invertibility of the QFT is inherited from the complex case. So, one sufficient (though not necessary) condition is that if $f(t)$ is absolutely integrable (i.e. $f(t) \in L^1(\mathbb{R}; \mathbb{H})$) and if it is of bounded variation on every finite interval, then $F^L(\omega)$ (and $F^R(\omega)$) exists and $f(t)$ can be recovered from the inverse Fourier transform relationship at each point at which $f(t)$ is continuous.

Since $F_s(\omega), F_p(\omega) \in \mathbb{C}_{\boldsymbol{\mu}}$, they can be computed using the *complex* Fourier transforms of $f_s(t)$ and $f_p(t)$. This is done by replacing $\boldsymbol{\mu}$ by I in f_s and f_p, taking their complex Fourier transforms, and then back-substituting $\boldsymbol{\mu}$ for I in the results. This is also the basis of a numerical implementation method using complex FFTs, which is discussed in section 3.3.1.3.

3.1.2. *Basic transform pairs*

There are several basic relationships that are of use in manipulating Fourier pairs, summarized in Table 3.2. For the most part, their proofs are elementary. The art of determining the Fourier transform of a function is in using these relationships.

Linearity: of the transform integrals implies

$$\alpha f_1(t) + \beta f_2(t) \rightleftharpoons \alpha F_1(\omega) + \beta F_2(\omega), \qquad [3.13]$$

where $\alpha, \beta \in \mathbb{C}_{\boldsymbol{\mu}}$ and $f_n(t) \rightleftharpoons F_n(\omega)$. Hence, all four 1D QFT definitions are linear operators.

Scaling: either time or frequency results in the following pairs:

$$f(\alpha t) \rightleftharpoons \frac{1}{|\alpha|} F\left(\frac{\omega}{\alpha}\right) \quad \text{and} \quad F(\alpha\omega) \rightleftharpoons \frac{1}{|\alpha|} f\left(\frac{t}{\alpha}\right), \qquad [3.14]$$

where $\alpha \in \mathbb{R}$.

Shifting: The Fourier pairs for shifts, in either time $f(t-\tau)$ or frequency $F(\omega-\boldsymbol{\nu})$, are unique for each Fourier definition, namely:

$$f(t-\tau) \rightleftharpoons F^R_{-\boldsymbol{\mu}}(\omega)\, e^{-\boldsymbol{\mu}\omega\tau} \quad \text{and} \quad F^R_{-\boldsymbol{\mu}}(\omega - \boldsymbol{\nu}) \rightleftharpoons f(t)\, e^{\boldsymbol{\mu}\boldsymbol{\nu}t} \qquad [3.15]$$

or

$$f(t-\tau) \rightleftharpoons e^{-\boldsymbol{\mu}\omega\tau}\, F^L_{-\boldsymbol{\mu}}(\omega) \quad \text{and} \quad F^L_{-\boldsymbol{\mu}}(\omega - \boldsymbol{\nu}) \rightleftharpoons e^{\boldsymbol{\mu}\boldsymbol{\nu}t}\, f(t) \qquad [3.16]$$

depending on the chirality of the transform used. As with the complex Fourier transforms, shifts and modulation operations are inter-related between the time and frequency domains, but the handedness of the transform must be taken into account.

Left-sided:	$\frac{1}{\sqrt{2\pi}}\int_{-\infty}^{+\infty} e^{\boldsymbol{\mu}\omega t} F(\omega)\,\mathrm{d}\omega = f(t)$	$\rightleftharpoons$	$F(\omega) = \frac{1}{\sqrt{2\pi}}\int_{-\infty}^{+\infty} e^{-\boldsymbol{\mu}\omega t} f(t)\,\mathrm{d}t$	
Right-sided:	$\frac{1}{\sqrt{2\pi}}\int_{-\infty}^{+\infty} F(\omega) e^{\boldsymbol{\mu}\omega t}\,\mathrm{d}\omega = f(t)$	$\rightleftharpoons$	$F(\omega) = \frac{1}{\sqrt{2\pi}}\int_{-\infty}^{+\infty} f(t) e^{-\boldsymbol{\mu}\omega t}\,\mathrm{d}t$	
Time-domain	$f(t)$	$\rightleftharpoons$	$F(\omega)$	Frequency-domain
Time reversal	$f(-t)$	$\rightleftharpoons$	$F(-\omega)$	Frequency reversal
Quaternion conjugation	$\overline{f(t)}$	$\rightleftharpoons$		See section 3.1.4
Reversed conjugation	$\overline{f(-t)}$	$\rightleftharpoons$		See section 3.1.4
Real	$f(t) \in \mathbb{R}$	$\rightleftharpoons$	$F(\omega) = \overline{F}(-\omega)$	Even, symmetric
Pure $\boldsymbol{\mu}$-imaginary	$f(t) \in \boldsymbol{\mu}\mathbb{R}$	$\rightleftharpoons$	$F(\omega) = -\overline{F}(-\omega)$	Odd, anti-symmetric
Even, symmetric	$f(t) = f^{\star}(-t)$	$\rightleftharpoons$	$F(\omega) \in \mathbb{R}$	Real
Odd, anti-symmetric	$f(t) = -f^{\star}(-t)$	$\rightleftharpoons$	$F(\omega) \in \boldsymbol{\mu}\mathbb{R}$	Pure $\boldsymbol{\mu}$-imaginary
Simplex	$f(t) \in \mathbb{C}_{\boldsymbol{\mu}_2}$	$\rightleftharpoons$	$F(-\omega) = \overline{F}(\omega)$	
Pure perplex	$f(t) \in \mathbb{C}_{\boldsymbol{\mu}_2}\boldsymbol{\mu}$	$\rightleftharpoons$	$F(-\omega) = -\overline{F}(\omega)$	
Time shift	$f(t-\tau)$	$\rightleftharpoons$	$e^{-\boldsymbol{\mu}\omega\tau} F(\omega)$	(Left-sided transform)
	$f(t-\tau)$	$\rightleftharpoons$	$F(\omega)\, e^{-\boldsymbol{\mu}\omega\tau}$	(Right-sided transform)
Modulation (left-sided transform)	$e^{\boldsymbol{\mu}\boldsymbol{\nu}t} f(t)$	$\rightleftharpoons$	$F(\omega - \boldsymbol{\nu})$	Frequency shift
Modulation (right-sided transform)	$f(t)\, e^{\boldsymbol{\mu}\boldsymbol{\nu}t}$	$\rightleftharpoons$	$F(\omega - \boldsymbol{\nu})$	Frequency shift
Time scale	$f(\alpha t)$	$\rightleftharpoons$	$\frac{1}{\lvert\alpha\rvert} F(\frac{\omega}{\alpha})$	
	$\frac{1}{\lvert\alpha\rvert} f(\frac{t}{\alpha})$	$\rightleftharpoons$	$F(\alpha\omega)$	Frequency scale
Linearity	$\alpha f_1(t) + \beta f_2(t)$	$\rightleftharpoons$	$\alpha F_1(\omega) + \beta F_2(\omega)$	
Convolution	$(f_1 * f_2)(t)$	$\rightleftharpoons$		See section 3.1.5
Correlation	$(f_1 \star f_2)(t)$	$\rightleftharpoons$		See section 3.1.5
Delta function	$\delta(t)\,\sqrt{2\pi}$	$\rightleftharpoons$	1	
Shifted delta	$\delta(t-\tau)\,\sqrt{2\pi}$	$\rightleftharpoons$	$e^{-\boldsymbol{\mu}\omega\tau}$	
	$e^{\boldsymbol{\mu}\boldsymbol{\nu}t}$	$\rightleftharpoons$	$\sqrt{2\pi}\,\delta(\omega - \boldsymbol{\nu})$	Shifted delta
	1	$\rightleftharpoons$	$\sqrt{2\pi}\,\delta(\omega)$	Delta function
nth derivative	$\frac{\mathrm{d}^n}{\mathrm{d}t^n} f(t)$	$\rightleftharpoons$	$(\boldsymbol{\mu}\omega)^n\, F(\omega)$	(Left-sided transform)
nth derivative	$\frac{\mathrm{d}^n}{\mathrm{d}t^n} f(t)$	$\rightleftharpoons$	$F(\omega)\,(\boldsymbol{\mu}\omega)^n$	(Right-sided transform)
(Left-sided transform)	$t^n f(t)$	$\rightleftharpoons$	$\boldsymbol{\mu}^n \frac{\mathrm{d}^n}{\mathrm{d}\omega^n} F(\omega)$	nth derivative
(Right-sided transform)	$t^n f(t)$	$\rightleftharpoons$	$\frac{\mathrm{d}^n}{\mathrm{d}\omega^n} F(\omega)\,\boldsymbol{\mu}^n$	nth derivative

NOTE.– $f(t) : \mathbb{R} \rightarrow \mathbb{H}$, $\boldsymbol{\mu}_2 \perp \boldsymbol{\mu}$, and $\alpha, \beta \in \mathbb{R}$.

Table 3.2. *1D forward quaternion Fourier transform pairs*

Derivatives: In a similar manner, derivatives, with respect to either time or frequency, have a chirality as:

$$\frac{\mathrm{d}^n}{\mathrm{d}t^n} f(t) \rightleftharpoons F^R_{-\boldsymbol{\mu}}(\omega)\,(\boldsymbol{\mu}\omega)^n \quad \text{and} \quad \frac{\mathrm{d}^n}{\mathrm{d}\omega^n} F^R_{-\boldsymbol{\mu}}(\omega)\,\boldsymbol{\mu}^n \rightleftharpoons t^n f(t) \qquad [3.17]$$

or

$$\frac{\mathrm{d}^n}{\mathrm{d}t^n} f(t) \rightleftharpoons (\boldsymbol{\mu}\omega)^n F^L_{-\boldsymbol{\mu}}(\omega) \quad \text{and} \quad \boldsymbol{\mu}^n \frac{\mathrm{d}^n}{\mathrm{d}\omega^n} F^L_{-\boldsymbol{\mu}}(\omega) \rightleftharpoons t^n f(t). \qquad [3.18]$$

Impulse functions: The QFT of the Dirac delta function $\delta(t)$ mimics the complex transform exactly and is identical for all QFT forms:

$$\delta(t)\,\sqrt{2\pi} \rightleftharpoons 1 \quad \text{and} \quad 1 \rightleftharpoons \sqrt{2\pi}\,\delta(\omega). \qquad [3.19]$$

Likewise, the shifted delta function is common across all QFT forms as:

$$\delta(t-\tau)\,\sqrt{2\pi} \rightleftharpoons e^{-\boldsymbol{\mu}\omega\tau} \quad \text{and} \quad e^{\boldsymbol{\mu}\boldsymbol{\nu}t} \rightleftharpoons \sqrt{2\pi}\,\delta(\omega-\boldsymbol{\nu}). \qquad [3.20]$$

In the last equation above, if the reverse transforms are used, then the sign of $\boldsymbol{\mu}$ is changed.

3.1.3. *Decompositions*

In the time domain, complex-valued functions $f(t)$ may happen to have one or more special symmetries. It might be purely real or purely imaginary, or it might be even, $f(t) = f(-t)$, or odd, $f(t) = -f(-t)$. These time-domain symmetries lead to symmetries in the frequency domain. Put another way, any function $f(t)$ can always be decomposed into even-odd or real-imaginary or both parts. Decomposition of 1D QFTs is helpful in simplifying QFT pairs and studying the relationships among the various definitions.

3.1.3.1. *Even-odd split*

In general, the function $f(t)$ is quaternion-valued; denoting by $f_s(t)$ and $f_p(t)$ its simplex and perplex parts with respect to the transform axis $\boldsymbol{\mu}$, we have[3]:

$$f(t) = f_s(t) + f_p(t)\boldsymbol{\mu}, \quad \text{where } f_s(t), f_p(t) \in \mathbb{C}_{\boldsymbol{\mu}_2}, \boldsymbol{\mu}_2 \perp \boldsymbol{\mu}, \qquad [3.21]$$

3 It should be noted that $\boldsymbol{\mu}_2 \perp \boldsymbol{\mu}$ allows for design freedom. If $\boldsymbol{\mu}_2$ is fixed, then the transform axis $\boldsymbol{\mu}$ can be steered as needed in the plane perpendicular to $\boldsymbol{\mu}_2$. Conversely, if $\boldsymbol{\mu}$ is fixed, then $\boldsymbol{\mu}_2$ can be steered in the plane perpendicular to $\boldsymbol{\mu}$.

and its (right-handed) QFT is

$$F(\omega) = F_s(\omega) + F_p(\omega)\boldsymbol{\mu}, \text{ where } F_s(\omega), F_p(\omega) \in \mathbb{C}_{\boldsymbol{\mu}_2} \quad [3.22]$$

where

$$S(\omega) = \quad \kappa_- \int_{\mathbb{R}} [f_s(t) \cos \omega t + f_p(t) \sin \omega t] \; \mathrm{d}t, \quad [3.23]$$

$$F_p(\omega) = -\kappa_- \int_{\mathbb{R}} [f_s(t) \sin \omega t - f_p(t) \cos \omega t] \; \mathrm{d}t. \quad [3.24]$$

Likewise, using the QFT inversion formula; with $F(\omega) = F_s(\omega) + F_p(\omega)\boldsymbol{\mu}$, we obtain:

$$f_s(t) = \kappa_+ \int_{\mathbb{R}} [F_s(\omega) \cos \omega t - F_p(\omega) \sin \omega t] \; \mathrm{d}\omega, \quad [3.25]$$

$$f_p(t) = \kappa_+ \int_{\mathbb{R}} [F_s(\omega) \sin \omega t + F_p(\omega) \cos \omega t] \; \mathrm{d}\omega \quad [3.26]$$

Now under the special condition that $f(t)$ is strictly $\boldsymbol{\mu}_2$-complex, i.e. $f_p(t) = 0$, then the simplex and perplex parts of $F(\omega)$ are given by:

$$F_s(\omega) = \kappa_- \int_{\mathbb{R}} f_s(t) \cos \omega t \, \mathrm{d}t \;\text{ and }\; F_p(\omega) = -\kappa_- \int_{\mathbb{R}} f_s(t) \sin \omega t \, \mathrm{d}t \quad [3.27]$$

From which we conclude that $F_s(\omega)$ is even and $F_p(\omega)$ is odd, i.e.

$$F_s(-\omega) = F_s(\omega) \;\text{ and }\; F_p(-\omega) = -F_p(\omega). \quad [3.28]$$

Therefore:

$$F^R_{-\boldsymbol{\mu}}(-\omega) = \overline{F^R_{-\boldsymbol{\mu}}(\omega)}, \;\text{ if }\; f(t) \in \mathbb{C}_{\boldsymbol{\mu}_2}, \; \boldsymbol{\mu}_2 \perp \boldsymbol{\mu}. \quad [3.29]$$

Conversely, if $F^R_{-\boldsymbol{\mu}}(-\omega) = \overline{F^R_{-\boldsymbol{\mu}}(\omega)}$, then $f_p(t) = 0$, i.e. $f(t) \in \mathbb{C}_{\boldsymbol{\mu}_2}$.

Starting instead with $f(t)$ being entirely perplex, i.e. $f_s(t) = 0$, so that $f(t) = f_p(t)\boldsymbol{\mu}$, then the simplex and perplex parts of $F(\omega)$ are given by

$$F_s(\omega) = \kappa_- \int_{\mathbb{R}} f_p(t) \sin \omega t \, \mathrm{d}t \;\text{ and }\; F_p(\omega) = \kappa_- \int_{\mathbb{R}} f_p(t) \cos \omega t \, \mathrm{d}t \quad [3.30]$$

Therefore, $F_s(\omega)$ is odd and $F_p(\omega)$ is even, i.e.

$$F_s(-\omega) = -F_s(\omega) \text{ and } F_p(-\omega) = F_p(\omega). \qquad [3.31]$$

Therefore:

$$F^R_{-\boldsymbol{\mu}}(-\omega) = -\overline{F^R_{-\boldsymbol{\mu}}(\omega)}, \text{ if } f(t) \notin \mathbb{C}_{\boldsymbol{\mu}_2}, \boldsymbol{\mu}_2 \perp \boldsymbol{\mu}. \qquad [3.32]$$

Conversely, if $F^R_{-\boldsymbol{\mu}}(-\omega) = -\overline{F^R_{-\boldsymbol{\mu}}(\omega)}$, then $f_s(t) = 0$, i.e. $f(t) \notin \mathbb{C}_{\boldsymbol{\mu}_2}$.

Imposing further restrictions on $\boldsymbol{\mu}_2$-complex $f(t)$, if we assume $f(t)$ is an even function of time, i.e. $f(-t) = f(t)$, then $F_p(\omega) = 0$. Conversely, if the QFT is a $\boldsymbol{\mu}_2$-complex simplex function, i.e. $F_p(\omega) = 0$, then $f(-t) = f(t)$, so that $f(t)$ is even. However, if we assume instead $f(t)$ is $\boldsymbol{\mu}_2$-complex and an odd function of time, i.e. $f(-t) = -f(t)$, then $F_s(\omega) = 0$. Conversely, if the QFT of a $\boldsymbol{\mu}_2$-complex function is perplex, i.e. $F_s(\omega) = 0$, then $f(-t) = -f(t)$, so that $f(t)$ is odd.

What we have shown is that if $f(t)$ is $\boldsymbol{\mu}_2$-complex and split into even and odd parts as:

$$f(t) = f_e(t) + f_o(t) \qquad [3.33]$$

where $f_e(t)$ is even with respect to time and $f_o(t)$ is odd, then:

$$F(\omega) = F_{se}(\omega) + F_{po}(\omega)\boldsymbol{\mu} \qquad [3.34]$$

where

$$F_{se}(\omega) = \mathcal{F}^R\{f_e\} \text{ and } F_{po}(\omega)\boldsymbol{\mu} = \mathcal{F}^R\{f_o\}. \qquad [3.35]$$

To switch to left-sided QFT's one instead starts with an alternate simplex-perplex decomposition, namely, let

$$f(t) = f_s(t) + \boldsymbol{\mu} f_p(t)$$

where the $\boldsymbol{\mu}$ factor is on the left of $f_p(t)$ instead of on the right [4].

The above analysis mimics the symmetry properties of the complex Fourier transform, but we speak of $\boldsymbol{\mu}_2$-complex even and odd functions instead of real even and odd functions of time.

4 Note this f_p is the conjugate of f_p from [3.21] by the swap rule.

It is desirable that the complex Fourier transform properties be contained within the QFT as a special case. Indeed this is so, for if we restrict the function to the $\boldsymbol{\mu}$-complex subfield of $\mathbb{H}$, then the classic complex properties carry over, namely, if $f(t)$ is strictly real-valued then:

$$F(\omega) = \overline{F}(-\omega)$$

and if $f(t)$ is purely $\boldsymbol{\mu}$-imaginary, i.e. $f(t) \in \boldsymbol{\mu}\mathbb{R}$,

$$F(\omega) = -\overline{F}(-\omega).$$

Likewise, if $f(t)$ is even, symmetric, i.e. $f(t) = \overline{f}(-t)$, then $F(\omega) \in \mathbb{R}$, but if $f(t)$ is odd, anti-symmetric $f(t) = -\overline{f}(-t)$, then $F(\omega) \in \boldsymbol{\mu}\mathbb{R}$. Proof of this is simple; since both $f(t)$ and the kernel of the transform $e^{\boldsymbol{\mu}\omega t}$ are from the same complex subfield, $\mathbb{C}_{\boldsymbol{\mu}} = \mathbb{R} + \boldsymbol{\mu}\mathbb{R}$, they are isomorphic to the standard complex numbers.

3.1.3.2. *Orthogonal 2D planes split*

Every $q \in \mathbb{H}$ can be split as $q = q_+ + q_-$ with $q_\pm = \frac{1}{2}(q \pm \boldsymbol{\mu}_1 q \boldsymbol{\mu}_2)$ as in [1.58] where $\boldsymbol{\mu}_1\boldsymbol{\mu}_2 = \boldsymbol{\mu}_3$ [HIT 07]. In terms of components, if $q = r_0 + \boldsymbol{\mu}_1 r_1 + \boldsymbol{\mu}_2 r_2 + \boldsymbol{\mu}_3 r_3$, then

$$q_\pm = \{(r_0 \pm r_3) + \boldsymbol{\mu}_1(r_1 \pm r_2)\}\,\frac{1 \pm \boldsymbol{\mu}_3}{2} = \frac{1 \pm \boldsymbol{\mu}_3}{2}\,\{(r_0 \pm r_3) - \boldsymbol{\mu}_2(r_1 \pm r_2)\}\,.$$

3.1.4. *Inter-relationships between definitions*

The left- and right-handed transforms given in [3.5] and [3.7] are related, as we might expect, and the operations that play a role in these relationships are: conjugation, time/frequency reversal and symplectic decomposition.

The operation with the simplest effect to consider is conjugation. In the classical complex case (when *complex* signals are being considered), conjugation and time/frequency reversal are connected: conjugation in one domain corresponds to conjugation and reversal in the other. For example, $\mathcal{F}^\star\{f(t)\} = \mathcal{F}\{f^\star(-t)\}$. In the quaternion case, however, there is a difference, and the reason is the different behavior of the quaternion conjugate compared to the complex conjugate, when applied to a product. In the complex case, the conjugate of a product is equal to the product of the conjugates: $(z_1 z_2)^\star = z_1^\star\, z_2^\star$. In the quaternion case, however, *order must be reversed*: $\overline{q_1 q_2} = \overline{q_2}\;\overline{q_1}$. The difference between the left- and right-sided transforms is the order of the signal and exponential; hence, it is not surprising that change of handedness and conjugation are connected. Figure 3.1 shows how transforms of opposite handedness are related with respect to conjugation.

Algebraically, the relationships can be expressed as:

$$\mathcal{F}^{R}_{\pm\mu}\{\overline{f}\} = \overline{\mathcal{F}^{L}_{\mp\mu}\{f\}} \quad \text{and} \quad \mathcal{F}^{L}_{\pm\mu}\{\overline{f}\} = \overline{\mathcal{F}^{R}_{\mp\mu}\{f\}}. \qquad [3.36]$$

(For simplicity, the figure omits the permutations of sign in the transform axis $\boldsymbol{\mu}$).

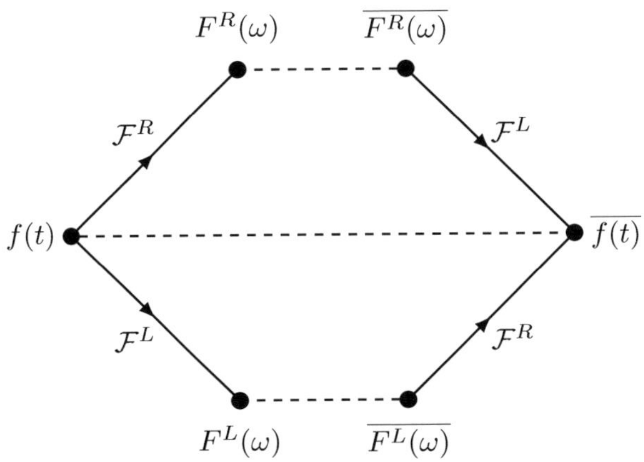

Figure 3.1. *Graph showing conjugation relationships between left- and right-handed one-sided QFTs. The dotted lines represent conjugation. The solid directed lines represent transforms*

Time/frequency reversal is slightly more complicated[5]. In the classical complex case, changing the sign of the time variable (time reversal) corresponds to changing the sign of the frequency variable (frequency reversal). Thus, if we have: $\mathcal{F}\{f(t)\} = F(\omega)$, then: $\mathcal{F}\{f(-t)\} = F(-\omega)$. In the quaternion case, however, the symplectic decomposition plays a part. Figure 3.2 shows how transforms of opposite handedness are related by time/frequency reversal and symplectic decomposition.

Note that application of the same transform twice to a signal $f(t)$ results in a time-reversed signal $f(-t)$ (ignoring scale factors). It is immaterial whether the transform chosen is left- or right-handed. The relationship between the two transforms is shown

5 Computationally time/frequency reversal is not trivial and we discuss it in section 3.3.3.2.

by the vertical dashed line that represents a *partial* frequency reversal. The symplectic decomposition of the right-handed QFT was given by [3.22] as:

$$F^R(\omega) = F_s(\omega) + F_p(\omega)\boldsymbol{\mu}, \qquad [3.37]$$

where $F_s(\omega), F_p(\omega) \in \mathbb{C}_{\boldsymbol{\mu}_2}$. The left-handed QFT using the same symplectic decomposition is found to be:

$$F^L(\omega) = F_s(\omega) + F_p(-\omega)\boldsymbol{\mu}, \qquad [3.38]$$

so that the simplex parts $F_s(\omega)$ are the same in both cases but the perplex parts $F_p(\pm\omega)$ of the two transforms are reversed in frequency. We have not considered the dual case, where the frequency-domain representation is reversed in frequency, but it should be obvious that the effect in the time domain would be a partial time reversal with only the perplex component reversed.

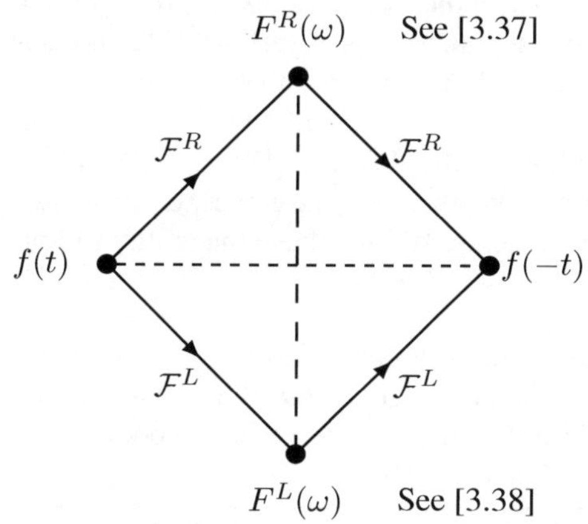

Figure 3.2. *Graph showing time/frequency reversal relationships between left- and right-handed one-sided QFTs. The dotted lines represent time/frequency reversal (horizontally) or partial time/frequency reversal (vertically). The solid directed lines represent transforms*

3.1.5. *Convolution and correlation theorems*

Theorems for convolution and correlation as shown in Table 3.1 for the complex Fourier transform are possible in the quaternion case, but the formulas are different for each formulation of transform. The results in the previous section show that left- and right-sided transforms are closely related, and conjugation plays a key role because of the interaction between conjugation and the re-ordering of a quaternion product, which are steps that often occur in the derivation of formulas involving quaternion products. Conjugation of a quaternion exponential requires a change of sign in the root of -1, and of course this corresponds to changing a forward transform into a reverse or *vice versa*. For these reasons, it is more sensible to construct such formulas using a mix of transforms, rather than confining the formula to a single transform, because this can yield a simpler formula. A secondary consideration is that when the forward and inverse transforms have different scale factors (as they do in some numerical implementations), combining forward and inverse transforms of the same handedness becomes awkward because of the different scale factors: it is simpler to combine transforms of the same direction (forward and inverse) of different handedness, in order to avoid such difficulties. We show how a correlation formula may be derived in section 4.2.2.

3.2. 2D quaternion Fourier transforms

The extension of the 1D QFT to two dimensions gives rise to a multitude of possibilities. The obvious choice of altering the transform kernel from $e^{-\boldsymbol{\mu}\omega t}$ to $e^{-\boldsymbol{\mu}(\omega x+\nu y)}$ for functions $f(x, y) : \mathbb{R} \to \mathbb{H}$ provides the same four left-right and forward-reverse transforms, but there are other choices. One concept left unexplored is the implication of the exponential product of two quaternions, i.e. in general $e^p e^q \neq e^{p+q}$, as noted in [1.38]. Hence, $e^{-(\boldsymbol{\mu}_1\omega x+\boldsymbol{\mu}_2\nu y)} \neq e^{-\boldsymbol{\mu}_1\omega x}e^{-\boldsymbol{\mu}_2\nu y}$. As soon as the exponent is split like this, we have the ability to "sandwich" the function $f(x, y)$ between the exponents. We also have the ability to use a different root of -1 for each exponent.

Table 3.3 details these eight forward forms. Doubling this number to account for the *reverse* transforms yields a total of 16 different QFTs. Only a limited number of most commonly used forms will be explored in this book.

	Left	Right	Sandwich
Single-axis	$e^{-\boldsymbol{\mu}(\omega x+\nu y)}\, f(.)$	$f(.)\, e^{-\boldsymbol{\mu}(\omega x+\nu y)}$	$e^{-\boldsymbol{\mu}\omega x}\, f(.)\, e^{-\boldsymbol{\mu}\nu y}$
Dual-axis	$e^{-(\boldsymbol{\mu}_1\omega x+\boldsymbol{\mu}_2\nu y)}\, f(.)$	$f(.)\, e^{-(\boldsymbol{\mu}_1\omega x+\boldsymbol{\mu}_2\nu y)}$	—
Factored	$e^{-\boldsymbol{\mu}_1\omega x}e^{-\boldsymbol{\mu}_2\nu y}\, f(.)$	$f(.)\, e^{-\boldsymbol{\mu}_1\omega x}e^{-\boldsymbol{\mu}_2\nu y}$	$e^{-\boldsymbol{\mu}_1\omega x}\, f(.)\, e^{-\boldsymbol{\mu}_2\nu y}$

Table 3.3. *Forward transform kernel definitions for* $f : \mathbb{R}^2 \to \mathbb{H}$

3.2.1. *Definitions*

The 2D single-axis, right-sided QFTs (forward and reverse) are defined as:

$$\mathcal{F}^R_{\mp\boldsymbol{\mu}}\{f\}(\omega, \nu) = F^R(\omega, \nu) = \int_{\mathbb{R}^2} f(x, y)\, e^{\mp\boldsymbol{\mu}(\omega x+\nu y)}\, \mathrm{d}x\, \mathrm{d}y, \qquad [3.39]$$

with their inverses being formed with the conjugate of the exponential kernel as:

$$\mathcal{F}_{\mp\boldsymbol{\mu}}^{-R}\left\{F^{R}\right\}(x,y) = f(x,y) = \int_{\mathbb{R}^2} F^{R}(\omega,\boldsymbol{\nu})\, e^{\pm\boldsymbol{\mu}(\omega x+\boldsymbol{\nu} y)}\, \mathrm{d}\omega\, \mathrm{d}\boldsymbol{\nu}, \qquad [3.40]$$

The 2D single-axis, left-sided QFTs (forward and reverse) are defined as:

$$\mathcal{F}_{\mp\boldsymbol{\mu}}^{L}\{f\}(\omega,\boldsymbol{\nu}) = F^{L}(\omega,\boldsymbol{\nu}) = \kappa_{\mp}^{2} \int_{\mathbb{R}^2} e^{\mp\boldsymbol{\mu}(\omega x+\boldsymbol{\nu} y)} f(x,y)\, \mathrm{d}x\, \mathrm{d}y, \qquad [3.41]$$

and their inverses are derived from the conjugate of the kernels as:

$$\mathcal{F}_{\mp\boldsymbol{\mu}}^{-L}\left\{F^{L}\right\}(x,y) = f(x,y) = \kappa_{\pm}^{2} \int_{\mathbb{R}^2} e^{\pm\boldsymbol{\mu}(\omega x+\boldsymbol{\nu} y)} F^{L}(\omega,\boldsymbol{\nu})\, \mathrm{d}\omega\, \mathrm{d}\boldsymbol{\nu}. \qquad [3.42]$$

The 2D factored, sandwich QFTs (forward and reverse) are defined as:

$$\mathcal{F}_{\mp(\boldsymbol{\mu}_1,\boldsymbol{\mu}_2)}^{S}\{f\}(\omega,\boldsymbol{\nu}) = F^{L}(\omega,\boldsymbol{\nu}) = \kappa_{\mp}^{2} \int_{\mathbb{R}^2} e^{\mp\boldsymbol{\mu}_1\omega x} f(x,y)\, e^{\mp\boldsymbol{\mu}_2\boldsymbol{\nu} y}\, \mathrm{d}x\, \mathrm{d}y, \qquad [3.43]$$

and their inverses are derived from the conjugate of the kernels as:

$$\mathcal{F}_{\mp(\boldsymbol{\mu}_1,\boldsymbol{\mu}_2)}^{-S}\left\{F^{L}\right\}(x,y) = f(x,y) = \kappa_{\pm}^{2} \int_{\mathbb{R}^2} e^{\pm\boldsymbol{\mu}_1\omega x} F^{S}(\omega,\boldsymbol{\nu})\, e^{\pm\boldsymbol{\mu}_2\boldsymbol{\nu} y}\, \mathrm{d}\omega\, \mathrm{d}\boldsymbol{\nu}. \qquad [3.44]$$

The remaining five transform definitions follow a similar pattern, with kernel placements as shown in Table 3.3.

Discovering which 2D quaternion-valued functions can be Fourier-transformed follows the same logic of the 1D case in section 3.1.1; we can borrow from the classic complex case. Split $f(x,y)$ via symplectic decomposition into simplex and perplex parts with respect to $\boldsymbol{\mu}_2$ as:

$$f(x,y) = f_s(x,y) + f_p(x,y)\,\boldsymbol{\mu}_2, \quad \text{where } f_s, f_p \in \mathbb{C}_{\boldsymbol{\mu}},$$

so that the right-sided QFT of [3.39] becomes:

$$\begin{aligned} F^R(\omega,\boldsymbol{\nu}) &= \int_{\mathbb{R}^2} f_s(x,y)\, e^{-\boldsymbol{\mu}(\omega x+\boldsymbol{\nu} y)}\, \mathrm{d}x\, \mathrm{d}y + \int_{\mathbb{R}^2} f_p(x,y)\, \boldsymbol{\mu}_2 e^{-\boldsymbol{\mu}(\omega x+\boldsymbol{\nu} y)}\, \mathrm{d}x\, \mathrm{d}y, \\ &= \int_{\mathbb{R}^2} f_s(x,y)\, e^{-\boldsymbol{\mu}(\omega x+\boldsymbol{\nu} y)}\, \mathrm{d}x\, \mathrm{d}y + \int_{\mathbb{R}^2} f_p(x,y)\, e^{\boldsymbol{\mu}(\omega x+\boldsymbol{\nu} y)}\, \mathrm{d}x\, \mathrm{d}y\, \boldsymbol{\mu}_2 \\ &= F_s(\omega,\boldsymbol{\nu}) + F_p(-\omega,-\boldsymbol{\nu})\boldsymbol{\mu}_2, \end{aligned}$$

In the same manner, [3.41] yields for the left-sided QFT:

$$\begin{aligned} F^L(\omega,\boldsymbol{\nu}) &= \int_{\mathbb{R}^2} e^{-\boldsymbol{\mu}(\omega x+\boldsymbol{\nu} y)} f_s(x,y)\, \mathrm{d}x\, \mathrm{d}y + \int_{\mathbb{R}^2} e^{-\boldsymbol{\mu}(\omega x+\boldsymbol{\nu} y)} f_p(x,y)\, \mathrm{d}x\, \mathrm{d}y\, \boldsymbol{\mu}_2 \\ &= F_s(\omega,\boldsymbol{\nu}) + F_p(\omega,\boldsymbol{\nu})\boldsymbol{\mu}_2 \end{aligned}$$

From the fact that the left- and right-sided 2D QFTs can be decomposed into a sum of complex subfield functions, $f_{s,p}(x,y)$, the existence and invertibility of these 2D QFTs is inherited from the 2D complex case. This is true even for the sandwich form of the QFT of [3.43], but showing this takes more work. Split $f(x,y)$ via symplectic decomposition into simplex and perplex parts with respect to $\boldsymbol{\mu}_2$ as:

$$f(x,y) = f_s(x,y) + f_p(x,y)\, \boldsymbol{\mu}_2, \quad \text{where } f_s, f_p \in \mathbb{C}_{\boldsymbol{\mu}_1},$$

so that [3.43] becomes:

$$F(\omega,\boldsymbol{\nu}) = F_s(\omega,\boldsymbol{\nu}) + F_p(\omega,\boldsymbol{\nu})\, \boldsymbol{\mu}_2 \qquad [3.45]$$

where

$$\begin{aligned} F_s(\omega,\boldsymbol{\nu}) &= \int_{\mathbb{R}^2} e^{-\boldsymbol{\mu}_1\omega x} f_s(x,y) e^{-\boldsymbol{\mu}_2\boldsymbol{\nu} y}\, \mathrm{d}x\, \mathrm{d}y, \text{ and} \\ F_p(\omega,\boldsymbol{\nu}) &= \int_{\mathbb{R}^2} e^{-\boldsymbol{\mu}_1\omega x} f_p(x,y) e^{-\boldsymbol{\mu}_2\boldsymbol{\nu} y}\, \mathrm{d}x\, \mathrm{d}y. \end{aligned}$$

The derivation begins with a 2D QFT equivalent to a 2D *complex* Fourier transform. Let:

$$F_{s,\boldsymbol{\mu}_1}(\omega,\boldsymbol{\nu}) = \int_{\mathbb{R}^2} e^{-\boldsymbol{\mu}_1\omega x} f_s(x,y) e^{-\boldsymbol{\mu}_1\boldsymbol{\nu} y}\, \mathrm{d}x\, \mathrm{d}y \qquad [3.46]$$

so that $F_{s,\boldsymbol{\mu}_1}(\omega,\boldsymbol{\nu}) \in \mathbb{C}_{\boldsymbol{\mu}_1}$ because both kernels of the QFT are the same. This is different from the definition of [3.44] on purpose. Expand the following expression, using Euler's cosine equation [1.51] in the last step, as:

$$
\begin{aligned}
&\tfrac{1}{2}\left\{F_{s,\boldsymbol{\mu}_1}(\omega,\boldsymbol{\nu}) + F_{s,\boldsymbol{\mu}_1}(\omega,-\boldsymbol{\nu})\right\} \\
&\quad = \tfrac{1}{2}\int_{\mathbb{R}^2} e^{-\boldsymbol{\mu}_1\omega x} f_s(x,y) e^{-\boldsymbol{\mu}_1\boldsymbol{\nu}y}\,\mathrm{d}x\,\mathrm{d}y + \tfrac{1}{2}\int_{\mathbb{R}^2} e^{-\boldsymbol{\mu}_1\omega x} f_s(x,y) e^{+\boldsymbol{\mu}_1\boldsymbol{\nu}y}\,\mathrm{d}x\,\mathrm{d}y \\
&\quad = \tfrac{1}{2}\int_{\mathbb{R}^2} e^{-\boldsymbol{\mu}_1\omega x} f_s(x,y)\left(e^{-\boldsymbol{\mu}_1\boldsymbol{\nu}y} + e^{+\boldsymbol{\mu}_1\boldsymbol{\nu}y}\right)\mathrm{d}x\,\mathrm{d}y \\
&\quad = \int_{\mathbb{R}^2} e^{-\boldsymbol{\mu}_1\omega x} f_s(x,y)\cos(\boldsymbol{\nu}y)\,\mathrm{d}x\,\mathrm{d}y
\end{aligned}
$$

Likewise, using the sine function of [1.51], another expression:

$$
\begin{aligned}
&\tfrac{1}{2}\left\{F_{s,\boldsymbol{\mu}_1}(\omega,\boldsymbol{\nu}) - F_{s,\boldsymbol{\mu}_1}(\omega,-\boldsymbol{\nu})\right\} \\
&\quad = \tfrac{1}{2}\int_{\mathbb{R}^2} e^{-\boldsymbol{\mu}_1\omega x} f_s(x,y) e^{-\boldsymbol{\mu}_1\boldsymbol{\nu}y}\,\mathrm{d}x\,\mathrm{d}y - \tfrac{1}{2}\int_{\mathbb{R}^2} e^{-\boldsymbol{\mu}_1\omega x} f_s(x,y) e^{+\boldsymbol{\mu}_1\boldsymbol{\nu}y}\,\mathrm{d}x\,\mathrm{d}y \\
&\quad = \tfrac{1}{2}\int_{\mathbb{R}^2} e^{-\boldsymbol{\mu}_1\omega x} f_s(x,y)\left(e^{-\boldsymbol{\mu}_1\boldsymbol{\nu}y} - e^{+\boldsymbol{\mu}_1\boldsymbol{\nu}y}\right)\mathrm{d}x\,\mathrm{d}y \\
&\quad = \left\{\int_{\mathbb{R}^2} e^{-\boldsymbol{\mu}_1\omega x} f_s(x,y)\sin(\boldsymbol{\nu}y)\,\mathrm{d}x\,\mathrm{d}y\right\}\{-\boldsymbol{\mu}_1\}
\end{aligned}
$$

Combining these two expressions, with the second post-multiplied by $-\boldsymbol{\mu}_3(=\boldsymbol{\mu}_2\boldsymbol{\mu}_1)$, yields:

$$
\begin{aligned}
&\tfrac{1}{2}\left\{F_{s,\boldsymbol{\mu}_1}(\omega,\boldsymbol{\nu}) + F_{s,\boldsymbol{\mu}_1}(\omega,-\boldsymbol{\nu})\right\} + \tfrac{1}{2}\left\{F_{s,\boldsymbol{\mu}_1}(\omega,\boldsymbol{\nu}) - F_{s,\boldsymbol{\mu}_1}(\omega,-\boldsymbol{\nu})\right\}\{-\boldsymbol{\mu}_3\} \\
&\quad = \int_{\mathbb{R}^2} e^{-\boldsymbol{\mu}_1\omega x} f_s(x,y) e^{-\boldsymbol{\mu}_2\boldsymbol{\nu}y}\,\mathrm{d}x\,\mathrm{d}y = F_s(\omega,\boldsymbol{\nu})
\end{aligned}
$$

which is the simplex part of the 2D QFT we are looking for from [3.45]. So that, after simplifying the combined expressions,

$$F_s(\omega,\boldsymbol{\nu}) = F_{s,\boldsymbol{\mu}_1}(\omega,\boldsymbol{\nu})\,\frac{1-\boldsymbol{\mu}_3}{2} + F_{s,\boldsymbol{\mu}_1}(\omega,-\boldsymbol{\nu})\,\frac{1+\boldsymbol{\mu}_3}{2} \qquad [3.47]$$

Substituting $f_p(x,y)$ for $f_s(x,y)$ in [3.46] provides similar expressions for the perplex part of the 2D QFT, namely:

$$F_p(\omega,\boldsymbol{\nu}) = F_{p,\boldsymbol{\mu}_1}(\omega,\boldsymbol{\nu})\,\frac{1-\boldsymbol{\mu}_3}{2} + F_{p,\boldsymbol{\mu}_1}(\omega,-\boldsymbol{\nu})\,\frac{1+\boldsymbol{\mu}_3}{2} \qquad [3.48]$$

which gives us the desired result:

$$\begin{aligned} F(\omega,\boldsymbol{\nu}) = &\ F_s(\omega,\boldsymbol{\nu}) + F_p(\omega,\boldsymbol{\nu})\,\boldsymbol{\mu}_2 \\ = &\ F_{s,\boldsymbol{\mu}_1}(\omega,\boldsymbol{\nu})\,\frac{1-\boldsymbol{\mu}_3}{2} + F_{s,\boldsymbol{\mu}_1}(\omega,-\boldsymbol{\nu})\,\frac{1+\boldsymbol{\mu}_3}{2} \\ &+ F_{p,\boldsymbol{\mu}_1}(\omega,\boldsymbol{\nu})\,\frac{\boldsymbol{\mu}_2+\boldsymbol{\mu}_1}{2} + F_{p,\boldsymbol{\mu}_1}(\omega,-\boldsymbol{\nu})\,\frac{\boldsymbol{\mu}_2-\boldsymbol{\mu}_1}{2}. \end{aligned}$$

3.2.2. *Basic transform pairs*

Following is a list of theorems that can be proved using the defining equations of section 3.2.1; it is assumed that all functions considered actually have Fourier transforms. For the sake of brevity, we consider only the forward versions of each form; the reverse form pairs will be obvious. Table 3.4 summarizes the results that follow.

Linearity: of the transform integrals implies:

$$\alpha f_1(x,y) + \beta f_2(x,y) \rightleftharpoons \alpha F_1(\omega,\boldsymbol{\nu}) + \beta F_2(\omega,\boldsymbol{\nu}), \qquad [3.49]$$

where $\alpha,\beta\in\mathbb{R}$ and $f_n(x,y)\rightleftharpoons F_n(\omega,\boldsymbol{\nu})$. Hence, all eight 2D QFT definitions shown in Table 3.3 are linear operators.

Scaling: Consider the function $f(\alpha x,\beta y)$, which is obtained by replacing (x,y) by $(\alpha x,\beta y)$, where $\alpha,\beta\in\mathbb{R}$. The QFT of this function is given as:

$$\mathcal{F}^S_{\boldsymbol{\mu}_1,\boldsymbol{\mu}_2}\{f(\alpha x,\beta y)\} = \frac{1}{|\alpha|\,|\beta|}F^S\left(\frac{\omega}{\alpha},\frac{\boldsymbol{\nu}}{\beta}\right). \qquad [3.50]$$

Change of Coordinates: 2D functions can undergo a change of coordinates in the function's domain, or index space. In general, a change of coordinates is a mapping: $(x,y)\to(\alpha x+\beta y,\gamma x+\delta y)$ where $\alpha,\beta,\gamma,\delta\in\mathbb{R}$ and $\alpha\delta-\beta\gamma\neq 0$. So, consider the function $f(\alpha x+\beta y,\gamma x+\delta y)$, which is obtained by a change of coordinates of the function $f(x,y)$. The QFT of this function is given as:

$$\begin{aligned} &\mathcal{F}^S_{\boldsymbol{\mu}_1,\boldsymbol{\mu}_2}\{f(\alpha x+\beta y,\gamma x+\delta y)\} = \\ &\frac{1}{\alpha\delta-\beta\gamma}\left\{\frac{1+\boldsymbol{\mu}_3}{2}F(\omega\hat{\delta}-\boldsymbol{\nu}\hat{\gamma},\boldsymbol{\nu}\hat{\alpha}+\omega\hat{\beta}) + \frac{1-\boldsymbol{\mu}_3}{2}F(\omega\hat{\delta}+\boldsymbol{\nu}\hat{\gamma},\boldsymbol{\nu}\hat{\alpha}-\omega\hat{\beta})\right\} \end{aligned} \qquad [3.51]$$

where $\boldsymbol{\mu}_3=\boldsymbol{\mu}_1\boldsymbol{\mu}_2, \hat{\alpha}=\frac{\alpha}{\alpha\delta-\beta\gamma}, \hat{\beta}=\frac{\beta}{\alpha\delta-\beta\gamma}, \hat{\gamma}=\frac{\gamma}{\alpha\delta-\beta\gamma}, \hat{\delta}=\frac{\delta}{\alpha\delta-\beta\gamma}$.

Shifts: If a function is shifted in both directions by a constant (δ_x, δ_y), the spectral amplitude remains the same but a linear term is added to the respective "phase" of each axis as:

$$\mathcal{F}^S_{\boldsymbol{\mu}_1,\boldsymbol{\mu}_2}\{f(x-\delta_x, y-\delta_y)\} = e^{-\boldsymbol{\mu}_1\omega\delta_x} F^S(\omega,\boldsymbol{\nu}) e^{-\boldsymbol{\mu}_2\boldsymbol{\nu}\delta_y}. \qquad [3.52]$$

Derivatives: Taking the *n*th partial derivative of a function with respect to x, we conclude that:

$$\mathcal{F}^S_{\boldsymbol{\mu}_1,\boldsymbol{\mu}_2}\left\{\frac{\partial^n}{\partial x^n} f(x,y)\right\} = (\boldsymbol{\mu}_1\omega)^n F^S(\omega,\boldsymbol{\nu}). \qquad [3.53]$$

Likewise, the *m*th partial derivative with respect to y gives:

$$\mathcal{F}^S_{\boldsymbol{\mu}_1,\boldsymbol{\mu}_2}\left\{\frac{\partial^n}{\partial y^n} f(x,y)\right\} = F^S(\omega,\boldsymbol{\nu})\,(\boldsymbol{\mu}_2\boldsymbol{\nu})^n \qquad [3.54]$$

and together they yield:

$$\mathcal{F}^S_{\boldsymbol{\mu}_1,\boldsymbol{\mu}_2}\left\{\frac{\partial^{n+m}}{\partial y^n \partial y^m} f(x,y)\right\} = (\boldsymbol{\mu}_1\omega)^n F^S(\omega,\boldsymbol{\nu})\,(\boldsymbol{\mu}_2\boldsymbol{\nu})^m. \qquad [3.55]$$

These formulas for the partial derivatives do not guarantee that the transforms exist; but if they do, the formulas will be as given.

Impulse functions: In applications we use functions that are singular, e.g., the Dirac delta function. To obtain the following results, we use this definition for the delta function: with $f(x, y)$ an arbitrary function, continuous at a given point (x_o, y_o), $\delta(x, y)$ is such that:

$$\int_{\mathbb{R}^2} \delta(x-x_o, y-y_o) f(x,y)\,\mathrm{d}x\,\mathrm{d}y = f(x_o, y_o).$$

It is assumed here that $\delta(x, y)$ is a real-valued function so as to commute with $f(x, y)$. The QFT of the 2D Dirac delta function is identical for all QFT forms, namely,

$$\delta(x,y)\,2\pi \rightleftharpoons 1 \quad \text{and} \quad 1 \rightleftharpoons 2\pi\,\delta(\omega,\boldsymbol{\nu}). \qquad [3.56]$$

However, the shifted delta function is only common across all factored and single-axis 2D QFT forms as:

$$\delta(x-x_o, y-y_o)\,2\pi \rightleftharpoons e^{-\boldsymbol{\mu}_1\omega x_o} e^{-\boldsymbol{\mu}_2\boldsymbol{\nu} y_o}$$

and

$$e^{\boldsymbol{\mu}_1 \omega_o x} e^{\boldsymbol{\mu}_2 \boldsymbol{\nu}_o y} \rightleftharpoons 2\pi\, \delta(\omega - \omega_o, \boldsymbol{\nu} - \boldsymbol{\nu}_o).$$

Whereas for the dual-axis (un-factored) form:

$$\delta(x - x_o, y - y_o)\, 2\pi \rightleftharpoons e^{-(\boldsymbol{\mu}_1 \omega x_o + \boldsymbol{\mu}_2 \boldsymbol{\nu} y_o)}$$

and

$$e^{\boldsymbol{\mu}_1 \omega_o x + \boldsymbol{\mu}_2 \boldsymbol{\nu}_o y} \rightleftharpoons 2\pi\, \delta(\omega - \omega_o, \boldsymbol{\nu} - \boldsymbol{\nu}_o).$$

3.2.3. *Decompositions*

Any 2D signal can be split into two different ways with regard to symmetry: either into *fully* even and odd components along the x- and y-axes as[6]:

$$f(x, y) = f_{ee}(x, y) + f_{oe}(x, y) + f_{eo}(x, y) + f_{oo}(x, y),$$

where f_{eo} denotes the part of f that is even with respect to x and odd with respect to y, and so forth, or the signal may be split into *partial* even and odd components as

$$f(x, y) = f_e(x, y) + f_o(x, y) \qquad [3.57]$$

$$= f_e(x, y) + f_o(x, y) \qquad [3.58]$$

where f_e denotes the part of f that is even with respect to x and no symmetry is assumed with respect to y, and f_o denotes the part of f that is odd with respect to y and no symmetry is assumed with respect to x, etc. The partial even-odd split is related to the full even-odd split as:

$$f_e(x, y) = \tfrac{1}{2}\{f(x, y) + f(-x, y)\} = f_{ee}(x, y) + f_{eo}(x, y),$$

$$f_o(x, y) = \tfrac{1}{2}\{f(x, y) - f(-x, y)\} = f_{oe}(x, y) + f_{oo}(x, y),$$

$$f_e(x, y) = \tfrac{1}{2}\{f(x, y) + f(x, -y)\} = f_{ee}(x, y) + f_{eo}(x, y),$$

$$f_o(x, y) = \tfrac{1}{2}\{f(x, y) - f(x, -y)\} = f_{oe}(x, y) + f_{oo}(x, y),$$

6 Recall that a function is even along the x-axis if $f(x) = f(-x)$ and odd if $f(x) = -f(-x)$. Hence, an arbitrary signal can be split using: $f_e(x) = \frac{1}{2}\{f(x) + f(-x)\}$ and $f_o(x) = \frac{1}{2}\{f(x) - f(-x)\}$, so that $f(x) = f_e(x) + f_o(x)$.

while the full even-odd split is given by:

$$f_{ee}(x,y) = \tfrac{1}{4}\{f(x,y) + f(-x,y) + f(x,-y) + f(-x,-y)\},$$
$$f_{eo}(x,y) = \tfrac{1}{4}\{f(x,y) + f(-x,y) - f(x,-y) - f(-x,-y)\},$$
$$f_{oe}(x,y) = \tfrac{1}{4}\{f(x,y) - f(-x,y) + f(x,-y) - f(-x,-y)\},$$
$$f_{oo}(x,y) = \tfrac{1}{4}\{f(x,y) - f(-x,y) - f(x,-y) + f(-x,-y)\}.$$

Using the 2D sandwich QFT [3.43], it can be shown that:

$$F^S(\omega,\boldsymbol{\nu}) = F^S_{ee}(\omega,\boldsymbol{\nu}) - \boldsymbol{\mu}_1 F^S_{oe}(\omega,\boldsymbol{\nu}) - F^S_{eo}(\omega,\boldsymbol{\nu})\boldsymbol{\mu}_2 + \boldsymbol{\mu}_1 F^S_{oo}(\omega,\boldsymbol{\nu})\boldsymbol{\mu}_2$$

where

$$F^S_{ee}(\omega,\boldsymbol{\nu}) = \mathcal{F}^S_{-(\boldsymbol{\mu}_1,\boldsymbol{\mu}_2)}\left\{f_{ee}\right\}(\omega,\boldsymbol{\nu}),$$
$$F^S_{oe}(\omega,\boldsymbol{\nu}) = \mathcal{F}^S_{-(\boldsymbol{\mu}_1,\boldsymbol{\mu}_2)}\left\{f_{oe}\right\}(\omega,\boldsymbol{\nu}),$$
$$F^S_{eo}(\omega,\boldsymbol{\nu}) = \mathcal{F}^S_{-(\boldsymbol{\mu}_1,\boldsymbol{\mu}_2)}\left\{f_{eo}\right\}(\omega,\boldsymbol{\nu}),$$
$$F^S_{oo}(\omega,\boldsymbol{\nu}) = \mathcal{F}^S_{-(\boldsymbol{\mu}_1,\boldsymbol{\mu}_2)}\left\{f_{oo}\right\}(\omega,\boldsymbol{\nu}).$$

3.2.4. *Inter-relationships between definitions*

Using the symplectic decomposition of $f(x,y)$ with respect to $\boldsymbol{\mu}_1$, i.e.

$$f(x,y) = f_s(x,y) + f_p(x,y)\boldsymbol{\mu}_2, \text{ where } f_{s,p} \in \mathbb{C}_{\boldsymbol{\mu}_1}$$

the right-sided QFT is given by:

$$F^R(\omega,\boldsymbol{\nu}) = F_s(\omega,\boldsymbol{\nu}) + F_p(\omega,\boldsymbol{\nu})\boldsymbol{\mu}_2, \qquad [3.59]$$

where $F_s(\omega,\boldsymbol{\nu}), F_p(\omega,\boldsymbol{\nu}) \in \mathbb{C}_{\boldsymbol{\mu}_1}$. The left-sided QFT, using the same symplectic decomposition for $f(x,y)$, is found to be:

$$F^L(\omega,\boldsymbol{\nu}) = F_s(\omega,\boldsymbol{\nu}) + F_p(-\omega,-\boldsymbol{\nu})\boldsymbol{\mu}_2, \qquad [3.60]$$

so the simplex parts of the left-side and right-side QFTs are the same, while the perplex parts of the two transforms are reversed in frequency.

	$\frac{1}{2\pi}\int_{\mathbb{R}^2} e^{\boldsymbol{\mu}_1\omega x} F(\omega,\boldsymbol{\nu}) e^{\boldsymbol{\mu}_2\boldsymbol{\nu} y}\,\mathrm{d}\omega\,\mathrm{d}\boldsymbol{\nu}$	$\rightleftharpoons$	$\frac{1}{2\pi}\int_{\mathbb{R}^2} e^{-\boldsymbol{\mu}_1\omega x} f(x,y) e^{-\boldsymbol{\mu}_2\boldsymbol{\nu} y}\,\mathrm{d}x\,\mathrm{d}y$	
Spatial-domain	$f(x,y)$	$\rightleftharpoons$	$F(\omega,\boldsymbol{\nu})$	Frequency-domain
Spatial reversal	$f(-x,y)$	$\rightleftharpoons$	$F(-\omega,\boldsymbol{\nu})$	Frequency reversal
	$f(x,-y)$	$\rightleftharpoons$	$F(\omega,-\boldsymbol{\nu})$	
Spatial shift	$f(x-x_o,y-y_o)$	$\rightleftharpoons$	$e^{-\boldsymbol{\mu}_1\omega x_o}\,F(\omega,\boldsymbol{\nu})\,e^{-\boldsymbol{\mu}_2\boldsymbol{\nu} y_o}$	
	$e^{\boldsymbol{\mu}_1\omega_o x}\,f(x,y)\,e^{\boldsymbol{\mu}_2\boldsymbol{\nu}_o y}$	$\rightleftharpoons$	$F(\omega-\omega_o,\boldsymbol{\nu}-\boldsymbol{\nu}_o)$	Frequency shift
Spatial scale	$f(\alpha x,\beta y)$	$\rightleftharpoons$	$\frac{1}{\lvert\alpha\rvert\,\lvert\beta\rvert}F(\frac{\omega}{\alpha},\frac{\boldsymbol{\nu}}{\beta})$	
	$\frac{1}{\lvert\alpha\rvert\,\lvert\beta\rvert}f(\frac{x}{\alpha},\frac{y}{\beta})$	$\rightleftharpoons$	$F(\alpha\omega,\beta\boldsymbol{\nu})$	Frequency scale
Coordinate change	$f(\alpha x+\beta y,\gamma x+\delta y)$	$\rightleftharpoons$	$\frac{1}{\alpha\delta-\beta\gamma}\{\frac{1+\boldsymbol{\mu}_3}{2}F(\omega\hat{\delta}-\boldsymbol{\nu}\hat{\gamma},\boldsymbol{\nu}\hat{\alpha}+\omega\hat{\beta})$ $+\frac{1-\boldsymbol{\mu}_3}{2}F(\omega\hat{\delta}+\boldsymbol{\nu}\hat{\gamma},\boldsymbol{\nu}\hat{\alpha}-\omega\hat{\beta})\}$	
Linearity	$\alpha f_1(x,y)+\beta f_2(x,y)$	$\rightleftharpoons$	$\alpha F_1(\omega,\boldsymbol{\nu})+\beta F_2(\omega,\boldsymbol{\nu})$	
Delta function	$\delta(x,y)\,2\pi$	$\rightleftharpoons$	1	
Shifted delta	$\delta(x-x_o,y-y_o)\,2\pi$	$\rightleftharpoons$	$e^{-\boldsymbol{\mu}_1\omega x_o}e^{-\boldsymbol{\mu}_2\boldsymbol{\nu} y_o}$	
	$e^{\boldsymbol{\mu}_1\omega_o x}e^{\boldsymbol{\mu}_2\boldsymbol{\nu}_o y}$	$\rightleftharpoons$	$2\pi\,\delta(\omega-\omega_o,\boldsymbol{\nu}-\boldsymbol{\nu}_o)$	Shifted delta
	1	$\rightleftharpoons$	$2\pi\,\delta(\omega,\boldsymbol{\nu})$	Delta function
$n th$ derivative	$\frac{\partial^n}{\partial x^n}f(x,y)$	$\rightleftharpoons$	$(\boldsymbol{\mu}_1\omega)^n\,F(\omega,\boldsymbol{\nu})$	
	$\frac{\partial^n}{\partial y^n}f(x,y)$	$\rightleftharpoons$	$F(\omega,\boldsymbol{\nu})\,(\boldsymbol{\mu}_2\boldsymbol{\nu})^n$	
	$\frac{\partial^{n+m}}{\partial y^n\partial y^m}f(x,y)$	$\rightleftharpoons$	$(\boldsymbol{\mu}_1\omega)^n\,F(\omega,\boldsymbol{\nu})\,(\boldsymbol{\mu}_2\boldsymbol{\nu})^m$	

NOTE.– $f(t):\mathbb{R}^2\to\mathbb{H},\quad \alpha,\beta,\gamma,\delta\in\mathbb{R},\quad \boldsymbol{\mu}_3=\boldsymbol{\mu}_1\boldsymbol{\mu}_2$
$\hat{\alpha}=\frac{\alpha}{\alpha\delta-\beta\gamma},\hat{\beta}=\frac{\beta}{\alpha\delta-\beta\gamma},\hat{\gamma}=\frac{\gamma}{\alpha\delta-\beta\gamma},\hat{\delta}=\frac{\delta}{\alpha\delta-\beta\gamma}.$

Table 3.4. *2D factored, sandwich quaternion Fourier transform pairs*

If the signal $f(x,y)$ is split into the partial even and odd parts with respect to the y-axis, i.e. $f(x,y)=f_e(x,y)+f_o(x,y)$, then we find:

$$F^S(\omega,\boldsymbol{\nu})=F_e^L(\omega,\boldsymbol{\nu})-\boldsymbol{\mu}_1F_o^L(\omega,\boldsymbol{\nu})\boldsymbol{\mu}_2 \qquad [3.61]$$

where

$$F_e^L(\omega,\boldsymbol{\nu})=\mathcal{F}^L_{-\boldsymbol{\mu}}\{f_e\}(\omega,\boldsymbol{\nu})$$

$$F_o^L(\omega,\boldsymbol{\nu})=\mathcal{F}^L_{-\boldsymbol{\mu}}\{f_o\}(\omega,\boldsymbol{\nu})$$

so that the sandwich form can be constructed from the left-handed form. See [YEH 08] for details of the proof.

Likewise, if the same signal $f(x,y)$ is split into the partial even and odd parts with respect to the x-axis, i.e., $f(x,y)=f_e(x,y)+f_o(x,y)$, then we find:

$$F^S(\omega,\boldsymbol{\nu})=F_e^R(\omega,\boldsymbol{\nu})-\boldsymbol{\mu}_1F_o^R(\omega,\boldsymbol{\nu})\boldsymbol{\mu}_2 \qquad [3.62]$$

where

$$F_e^R(\omega, \boldsymbol{\nu}) = \mathcal{F}_{-\boldsymbol{\mu}}^R \{f_e\} (\omega, \boldsymbol{\nu})$$

$$F_o^R(\omega, \boldsymbol{\nu}) = \mathcal{F}_{-\boldsymbol{\mu}}^R \{f_o\} (\omega, \boldsymbol{\nu})$$

so that the sandwich form can also be constructed from the right-handed form.

3.3. Computational aspects

In this section, we consider the practical and numerical issues that arise in the computation of QFTs, and the process of verification that the numerical results computed by the transform are in agreement with the mathematical definition. Verification is also applicable to formulas such as those presented in section 3.1.4, and we discuss the details of time/frequency reversal, because this plays a key role in the inter-relationships between transforms.

3.3.1. *Coding*

Coding of a QFT requires at least some minimal quaternion arithmetic operations to be implemented. These operations are normally provided by a software library, and unless a library already exists for the chosen computing platform or language, it will be necessary to create one. Examples of libraries are the Boost library for C++, which contains some quaternion functions in the Math toolkit [BRI 13], and the QTFM library for MATLAB® [SAN 13b], which implements a wide range of quaternion functions using as far as possible standard MATLAB® notation and conventions. Both of these libraries are open source; so implementation of a quaternion library for another language or platform should not be difficult, since the code of the existing libraries can be inspected for ideas.

Even if a library exists, there may be bugs in it that render the results of a quaternion transform computation incorrect under some circumstances, and this fact should be considered when testing an implementation to verify correctness.

A further use for a quaternion library is the analysis of the results from computing a transform, whether this is some form of processing (e.g. filtering in the frequency domain) or interpretation (e.g. by plotting the Fourier coefficients or some aspects of them such as their magnitude, axis or angle).

In the following sections, we consider four possible ways to compute a QFT numerically:

1) a direct quaternion computation using a discrete QFT implementing directly the desired transform as a single or double summation;

2) a direct quaternion FFT algorithm using decimation in time or frequency or another method [BRA 00];

3) a decomposition into a pair of complex Fourier transforms that can be computed using existing complex FFT library routines;

4) a matrix method that avoids the need for a quaternion library, and thus provides a way to validate a computation based on a quaternion library.

Of these methods, the third is probably the best choice for most purposes, since it requires minimal effort to code, and exploits the significant work that others have invested in the fast and accurate computation of complex FFTs. We discuss each of the above methods in turn in the following sections.

3.3.1.1. *Direct computation of a DFT*

Any discrete Fourier transform (DFT) (complex, quaternion or otherwise) can be computed directly from the definition, using a summation (or double summation in the case of a 2D transform). This is often referred to in the literature as a DFT, the word *discrete* referring to the discretization of time (or position in an image). For example, a 1D left-sided QFT pair as defined in [3.7] could be computed in discrete form using the following pair of summations[7]:

$$\begin{aligned} F[u] &= \sum_{n=0}^{N-1} \exp\left(-\boldsymbol{\mu}\, 2\pi \frac{nu}{N}\right) f[n] \\ f[n] &= \frac{1}{N} \sum_{u=0}^{N-1} \exp\left(+\boldsymbol{\mu}\, 2\pi \frac{nu}{N}\right) F[u] \end{aligned} \qquad [3.63]$$

Now obviously the conversion from the integral notation to the discrete notation assumes another difference: the integral is over all time, whereas the summation is over a limited range of sample indices. This is a matter that is standard in digital signal and image processing, where the integral notation is often used, even though a discretized form is actually used for numerical computation.

The summation is simple enough to be coded in a few lines of code in a high-level language. In MATLAB®, it is possible to avoid a loop and perform a vectorized computation by precomputing the exponential for all index values[8], then multiply the

7 Exactly how the scale factor is handled is a matter of choice. It could be applied to the forward transform, or the square root of the scale factor could be applied in both directions.

8 This assumes there is enough memory available to store the entire set of exponential values. For longer transforms, this is not a reasonable assumption, and a more sophisticated computation is recommended in order to store the exponential values more economically.

resulting matrix by the signal vector to obtain the result. This is how the QTFM function `qdft.m` is coded, and the following code extract shows the core of the code:

```
F = E * f;
```

The variable E here contains a matrix of dimension $N \times N$ containing the values of the exponential function for all possible index values n and u in [3.63], as discussed, for example, by Golub and van Loan [GOL 96, section 4.6.4]. A single matrix product then multiplies the signal by this matrix to yield the frequency-domain coefficients. This works because the quaternion operations for element-wise and matrix multiplication are implemented within the toolbox, and the exponential function is vectorized, just like the standard MATLAB® exponential function. (Of course the classical complex DFT can be computed by exactly the same code, by replacing `mu` with `1i`.)

A major disadvantage of this method, whether implemented using loops, or as vectorized code, is that the run-time scales as the square of the length of the signal array, N. This can be easily seen in Figure 3.3, which was obtained by running the `qdft` function on random quaternion arrays from length 1 to 2,000, and averaging over three measurements. For comparison, it also shows the results from the `qfft` function (see section 3.3.1.2).

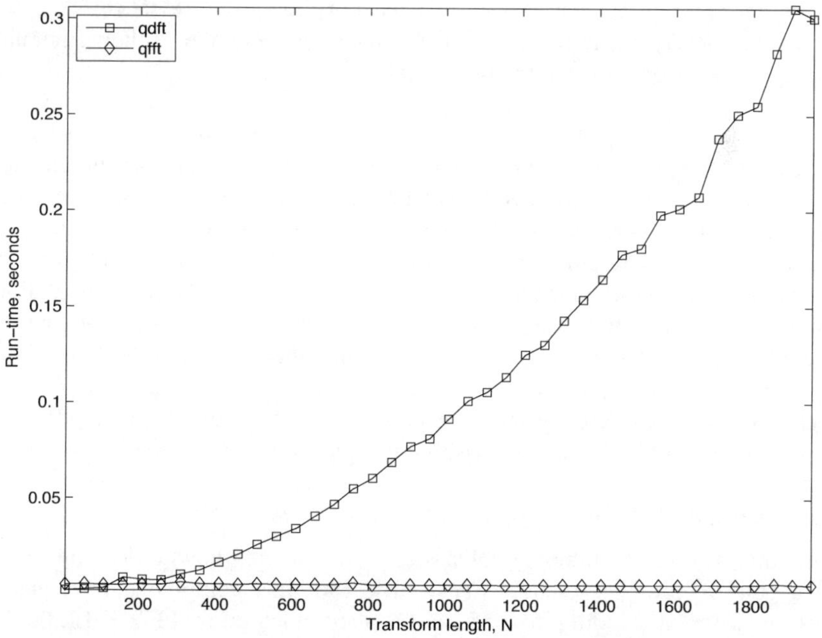

Figure 3.3. *Run-time for the `qdft` and `qfft` functions of the QTFM toolbox [SAN 13b] on random quaternion signals*

As can be seen, even for a short array of only 2,000 samples, the run-time for the `qdft` function is around a third of a second, whereas the `qfft` computes in approximately 10 ms, with a very much slower increase in run-time with increasing N. The corresponding transform for a 2D array has run-time that scales as N^4, assuming a square array of dimension $N \times N$. (None of this is unique to QFTs, but since computing with quaternions takes four times as much memory and 16 times as much time as computing with reals, run-time is an important issue.)

Despite the disadvantage of long run-times, and run-times that scale with the square of the signal length, a direct coding of a quaternion DFT is useful as a reference against which the results of faster methods can be compared. As can be seen, the translation of mathematics into code can be so direct that there is little room for error. However, as already explained, it is not difficult to verify the results of a quaternion DFT against a complex Fourier transform implementation.

3.3.1.2. *Direct coding of an FFT*

The fast Fourier transform (FFT) is a family of algorithms for computing the DFT with much reduced run-time scaling and better accuracy. Both advantages arise from a significant reduction in the number of arithmetic operations needed. We will not discuss here how this reduction in run-time is achieved, since it is not pertinent to this book, and is well-covered elsewhere (see, for example, [BRA 00]). A DFT takes time proportional to N^2, whereas an FFT has run-time proportional to $N \log N$. The difference is clearly illustrated in Figure 3.3 where the second trace shows the run-time of the `qfft` function (close to the horizontal axis).

Direct coding of a quaternion FFT algorithm is possible, but it requires significant effort, even for a power-of-two FFT (the easiest type of algorithm to implement), and it does not seem worth attempting nowadays. It was used by Sangwine in 1996 [SAN 96] because there was no other fast method capable of working with 512×512 color images, but it required very significant effort to code and test, even after a quaternion arithmetic library had been written (in Ada 95). The implementation was limited to array sizes that were powers of 2 because handling other sizes was much more difficult, and it was not necessary to handle other sizes in order to demonstrate the validity of using QFTs on color images. Direct coding of quaternion FFTs was effectively made obsolete by the idea presented in the next section, which is the method recommended for anyone wishing to code a quaternion FFT.

3.3.1.3. *Computation using decompositions and complex FFTs*

In 2000, two of the authors published a paper showing how to decompose a QFT into a pair of complex Fourier transforms, and thus compute a fast quaternion transform without needing to code an explicit quaternion FFT [ELL 00a]. The method is also explained in [ELL 07c, section IV.B]. This technique has a major advantage over an explicit quaternion FFT: it exploits the capabilities of *any existing* fast complex FFT code, and any future enhancements made to the code (subject to recompilation or relinking with the newer version, of course). The most obvious candidate is the excellent FFTW library written at MIT by Matteo Frigo and Steven G. Johnson [FRI 05][9]. FFTW is written in C, and it handles arbitrarily sized arrays of data, even with large prime-number lengths. For repeated calculation on multiple data of the same length, it may be fine-tuned to optimize the speed on a particular machine architecture and at a particular address in memory (the latter affects cache hits and misses, which impacts performance, and can be taken into account in the tuning process).

The mathematical basis of the decomposition method is described in section 1.4.1.4 where it is shown that any quaternion can be decomposed into two quaternions parallel and perpendicular to a chosen unit pure quaternion. Applying this idea to a particular QFT formulation, as in section 3.1.1 (see [3.9] and [3.11]), it is possible to decompose the signal and the transform into a pair of quaternion transforms each of which is isomorphic to a complex Fourier transform. Note that it is necessary to find the decomposition formula for each particular transform. After computing the pair of complex transforms, the complex results are used to construct isomorphic quaternions, which are then combined into the quaternion result. Numerically, a simpler implementation of this idea is to compute a change of basis of the vector part of the quaternion signal, and this is how the technique is actually implemented in the QTFM library [SAN 13b, Function `change_basis`]. Given a basis, for example, $\{1, \boldsymbol{\mu}_1, \boldsymbol{\mu}_2, \boldsymbol{\mu}_1\boldsymbol{\mu}_2\}$ and a quaternion $q = w + x\boldsymbol{i} + y\boldsymbol{j} + z\boldsymbol{k}$ expressed in the standard basis $\{1, \boldsymbol{i}, \boldsymbol{j}, \boldsymbol{k}\}$, the four numerical components of the quaternion expressed in the new basis are simply computed as:

$$
\begin{aligned}
w' &= w \\
x' &= \langle q, \boldsymbol{\mu}_1 \rangle \\
y' &= \langle q, \boldsymbol{\mu}_2 \rangle \\
z' &= \langle q, \boldsymbol{\mu}_3 \rangle
\end{aligned}
$$

After the change of basis, the numerical components of the quaternion signal can be directly utilized as complex signals $z_1 = w' + x'I$ and $z_2 = y' + z'I$, and the complex transforms computed. Finally, the change of basis has to be inverted after reassembling the complex results into a quaternion result. Inversion requires that the standard basis be expressed in terms of the new basis. This is straightforward because a basis can be expressed as an orthogonal 3×3 matrix, which is trivially invertible (note that the change of basis does not affect the scalar part). Therefore, the three unit quaternions needed to invert the change of basis are simply computed from $\boldsymbol{\mu}_1$, $\boldsymbol{\mu}_2$ and

9 FFTW is the library used by MATLAB® to compute FFTs. It is available at: http://www.fftw.org.

$\mu_1\mu_2$ by constructing new unit pure quaternions from their x, y and z components, in the same manner as constructing new rows from the columns of the corresponding 3×3 matrix.

3.3.2. *Verification*

The transforms presented in this book are not just theoretical results. They have been implemented in computer code and applied to problems in signal and image processing. This requires numerical implementations, and numerical implementations should be verified carefully. In the experience of the authors, numerical implementations also provide an excellent way to verify the mathematics, particularly with hypercomplex formulas where the interactions between non-commutativity and operations such as conjugation provide plentiful sources of error in mathematical derivations.

3.3.3. *Verification of transforms*

Here, we consider the problem of verifying that the numerical results produced by a transform implementation are in agreement with the mathematical definition. This is not a trivial issue: it is simple to check that an implementation of a particular transform inverts correctly – compute the forward transform of a randomly created signal or image f, giving F, and then compute the inverse transform of F, giving g. Then, f and g should be identical to within a suitably small tolerance (computing their difference is the simplest way to check this, followed by finding the maximum absolute value of the result). However, this process does not verify that the transform computes the Fourier coefficients correctly, according to the mathematical definition, although it is a useful initial check – if the test described fails (the difference between f and g is not small), then there is obviously an error in the coding of the transform. A most likely error is that f and g could differ significantly by a scale factor, which is not serious, and easily corrected. This problem is easily diagnosed by computing the quaternion *ratio* of the elements in f to those in g (multiply each sample in f by the quaternion inverse of the corresponding sample in g). A scale factor error will manifest as a constant real-valued result for all elements of f and g. If the error is not a scale factor error, the problem of diagnosis is more difficult. This is where it is valuable to code a DFT implementation, as discussed in section 3.3.1.2, because this provides a simple, directly coded implementation that is useful as a reference for a more complex implementation based on decompositions and complex FFTs. However, even a DFT implementation requires a quaternion library, and there could be errors in a library that affect both transform implementations. Given the limited number of hypercomplex libraries in existence, running on different platforms, and written in different languages, even checking one against another is a non-trivial task. Until 2010, there was no known solution for this problem, other than the use of a different library for verification, but the formulation presented in the next section even overcame that problem.

It is also possible to verify results against a classical complex Fourier transform implementation. The latter is possible because a QFT reduces to the classical complex Fourier transform under certain conditions. If these conditions are satisfied, a numerical comparison is possible between a quaternion implementation and a classical complex implementation. The most general way to do this is to construct a quaternion signal so that all the samples will have the same axis, and this axis is the same as the transform axis[10]. Here is the process: construct a complex test signal f (e.g. using a random number generator), such that $f = f_r + I f_i$. Then construct an isomorphic quaternion signal $q = f_r + \boldsymbol{\mu} f_i$. Compute the complex Fourier transform of the complex signal, $F = \mathcal{F}\{f\}$, and the QFT of the quaternion signal, with axis $\boldsymbol{\mu}$, $Q = \mathcal{F}\{q\}$. Then, to within a tolerance, we should find that $F_r = \mathrm{S}(Q)$ and $\boldsymbol{\mu} F_i = \mathbf{V}(Q)$.

3.3.3.1. *Transform formulation using a matrix exponential*

A solution to the problem of verifying a transform without requiring a second quaternion library was first published in 2012 [SAN 12], and it is of considerable mathematical interest. It is based on a matrix representation of quaternions (as already discussed in sections I.6.2 and 1.4.3), and it can compute a QFT *without a quaternion library*.

A one-sided quaternion DFT pair such as that given in [3.5] can be represented in the following form, which combines matrix representations of quaternions with quaternions represented as four-element column vectors:

$$F[u] = S \sum_{n=0}^{N-1} \exp\left(-\mathsf{J}\, 2\pi \frac{nu}{N}\right) f[n] \qquad [3.64]$$

$$f[n] = T \sum_{u=0}^{N-1} \exp\left(+\mathsf{J}\, 2\pi \frac{nu}{N}\right) F[u] \qquad [3.65]$$

where J is a 4×4 matrix root of -1 representing a quaternion root of -1 using the representation of [1.69]; $f[n]$ and $F[u]$ are quaternion-valued discrete-time signals with N samples; each sample, indexed by n or u respectively, being a column vector of four elements corresponding to the four components of the corresponding quaternion signal sample; and the two scale factors S and T multiply to give $1/N$ in the usual way (the distribution of the scale factors between the forward and inverse transforms being arbitrary).

10 It is assumed here that we are discussing a one-sided transform with a single axis. Obviously, matters get a bit more complicated if the transform is two-sided with multiple axes.

This formulation depends on two results that are not obvious at first glance. The first is that Euler's formula is valid when the root of -1 is replaced by a matrix that squares to a negated identity matrix (we call this a matrix root of -1): $\exp J\theta = \mathsf{I}\cos\theta + \mathsf{J}\sin\theta$ where I is an identity matrix, and $\mathsf{J}^2 = -\mathsf{I}$. For a proof, we refer the reader to [SAN 12], but note that the result is easily verified using MATLAB® or other mathematical software. The exponential, of course, must be a *matrix exponential* [ROW 14].

The second result is due to Ickes [ICK 70], who showed that the matrix representation of a quaternion, as discussed in section I.6.2 and 1.4.3 could be combined with a four-element column vector containing the four components of a quaternion to represent the quaternion product as a matrix-vector product (the result of course is a four-element column vector). Surprisingly, Ickes also showed that the matrix does not have to represent the left operand in the product: by transposing the lower-right 3×3 part of the matrix, the matrix-vector product yields the result when the quaternion represented by the matrix is the right operand, despite the fact that the matrix remains the left operand in the matrix-vector product.

The transform pair in [3.64] can be implemented in MATLAB® code as follows (note the directness of the coding):

```
N = size(f, 2);
F = zeros(size(f));
for n = 0:N-1
  for u = 0:N-1
    F(:, u + 1) = F(:, u + 1) + ...
    expm(-J .* 2 .* pi .* n .* u ./ N) * f(:, n + 1);
  end
end
```

Here, `F` and `f` are arrays of four rows and N columns, representing a signal of N quaternion samples; and `J` is the quaternion root of -1 used in the transform, represented as a 4×4 matrix according to [1.69]. This code runs on standard MATLAB®, i.e. without needing any quaternion toolbox or library, and it has been verified to produce the same numerical results as the quaternion functions `qdft` and `qfft` in the QTFM library [SAN 13b].

The approach discussed here has been shown also to work for two-sided transforms such as that in [3.43], using the simple approach of representing all the quaternions (left and right exponentials, and the signal samples) as matrices. However, using the matrix-vector approach of Ickes [ICK 70], it should be possible to formulate even a two-sided transform as a matrix-matrix-vector product, placing the two exponential matrices on the left, but using the middle one to represent the right-side exponential in Ickes' partially transposed manner.

Finally, note that the matrix formulation also works in the classical complex case, and for other hypercomplex algebras such as Clifford algebras [SAN 12].

3.3.3.2. *Time and frequency reversal*

As we have seen in section 3.1.4, the various transform formulations are inter-related, and there are two operations that play a key role in the inter-relationships: conjugation and time/frequency reversal. Conjugation is computationally simple, but it is worth expanding a little on time/frequency reversal because verification of relationships involving time/frequency reversal is not as simple as it might appear. In the continuous-time case, time and frequency reversal simply require the time or frequency variable to be negated, and this is mathematically and algebraically trivial. In discrete time, however, some care is needed to implement this operation correctly, whether working algebraically, or with a numerical implementation. To understand this point, it is necessary to understand how the frequency-domain coefficients of a discrete-time Fourier transform are stored.

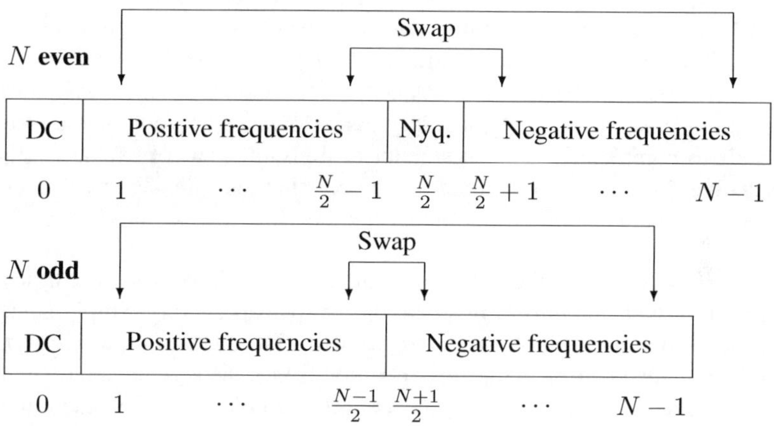

Figure 3.4. *Layout of frequency-domain coefficients for a DFT of N samples*

Figure 3.4 shows the layout of the samples in the frequency domain for a 1D DFT (classical or quaternion, the layout is the same). The indexing in the figure shows zero-based indexing, as used in mathematical formulas. In programming languages with one-based indexing, program code must be adapted to use array indices for accessing the transform elements, but zero-based indexing for calculations. In the even-length case, the coefficients are divided into four categories:

– the zero-frequency or DC coefficient, stored in the first index position with numerical index zero;

– the Nyquist frequency coefficient, stored in the first index position of the second half of the array with numerical index $N/2$;

– the positive frequency coefficients, stored between the DC and Nyquist coefficients, with the lowest frequency on the left, with index 1, and the highest frequency on the right, with index $(N/2) - 1$;

– the negative frequency coefficients, stored between the Nyquist coefficient and the end of the array, with the lowest frequency coefficients (i.e. closest to zero frequency) at the end, with index $N-1$, and the highest frequency next to the Nyquist coefficient, with coefficient $(N/2) + 1$.

In the odd-length case, the Nyquist frequency coefficient is not present. The zero-frequency coefficient is at index zero, the positive frequency coefficients run from index 1 to $(N - 1)/2$, and the negative frequency coefficients run from index $N - 1$ down to $(N + 1)/2$.

Time reversal of a signal corresponds to the interchanging of the positive and negative frequency coefficients. Classically, in digital signal processing texts concerned with real signals, time reversal corresponds to conjugation in the frequency domain (because the Fourier coefficients of a real signal have complex conjugate symmetry – the negative frequency coefficients are the conjugates of the corresponding positive frequency coefficients). However, as soon as we move from real signals to complex, it is much simpler to think in terms of exchanging positive and negative frequency coefficients, since this is fundamentally how the correspondence between the two domains operates.

The swaps indicated by the arrows in Figure 3.4 show how to achieve time reversal of a signal *in either domain*. In practice, the swaps can be very simply achieved by reversing the order of the whole array, excluding the DC coefficient, which should stay where it is in index position zero. This operation leaves the Nyquist coefficient (in the even case), unmoved. Alternatively, reversing the whole array, and then applying a right circular shift by one position, will put the DC coefficient back in index position zero. To understand why it is that the same operation works in both domains, note that in time, the sample at index zero must remain in the same position, since when negating the time variable, $t = 0$ remains at 0. Second, we must remember that the DFT treats the signal as periodically extended to infinity. Thus, the sample stored at the rightmost end of the array is also logically adjacent to the sample with index zero, but *on its left*.

In two dimensions, the above condition applies to both the rows and columns of an image. The simplest way to consider this is to imagine the reversal applied to the rows first, then to the columns of the result, and indeed this is a simple way to implement the operation numerically.

4

Signal and Image Processing

This chapter presents some examples of applications of quaternion Fourier transforms (QFTs) to process color images and complex-valued signals. The concepts presented in this chapter will be illustrated on simulated and real images and signals.

4.1. Generalized convolution

One of the early motivations for the study of QFTs is that they can be used to describe color image filters when the image pixels are treated as *vectors* and thus offer freedom to process color images holistically, rather than as separated color space components. This section illustrates the use of QFTs to process color images through the use of a vector convolution operation.

4.1.1. *Classical grayscale image convolution filters*

Convolution is a way of describing how systems change an input signal into an output signal. If the system under consideration is a linear time- (or shift-) invariant process, the Fourier domain description of that system is greatly simplified. Time invariance means that whether we apply an input to the system now or τ seconds from now, the system output will be identical except for a time delay of τ seconds. Hence, the system is time invariant because the output does not depend on the particular time at which the input is applied. A similar description applies to shift-invariant systems. For example, image processing filters can be described by their point-spread function, which is independent of where, in the input image, the filter is applied.

The classical 2D convolution of two functions, $f(x, y)$ and $h(x, y)$, is given as:

$$(f * h)(x, y) = \int_{-\infty}^{\infty} \int_{-\infty}^{\infty} f(x - x', y - y')\, h(x', y')\, \mathrm{d}x'\, \mathrm{d}y' \qquad [4.1]$$

where $f(x, y)$ is the input signal and $h(x, y)$ is the system's point-spread function. The discrete equivalent of this convolution, used in image processing, is given by:

$$(f * h)[n, m] = \sum_{i=0}^{N-1} \sum_{j=0}^{M-1} f[n - i, m - j]\, h[i, j] \qquad [4.2]$$

where $f[n, m] \in \mathbb{R}^{N \times M}$ is the input image, a scalar array of grayscale pixels. The filter convolution mask $h[n, m]$ is typically much smaller in dimensions than the input image. The 2D complex Fourier transform of this equation is given by

$$\mathcal{F}\{f * h\} = F[u, v]\, H[u, v] \qquad [4.3]$$

where $F[u, v] = \mathcal{F}\{f\}$ and $H[u, v] = \mathcal{F}\{h\}$. This is an element-wise product, not a matrix product. This equation, called the Fourier convolution theorem, gives insight into what the filter mask does to an arbitrary input image. It scales and phase shifts each unique frequency component of the input image, $F[u, v]$, for a specific frequency index $[u, v]$ by the spectral coefficient of the mask at the same index. In fact, the desired filter mask can be designed by construction of the mask spectral coefficients and taking the inverse transform of the results. Or, if only one specific image needs to be filtered, it can be manually shaped in the frequency domain and the result inverse transformed, skipping the construction of the mask altogether.

The discrete convolution of [4.2] can be visualized by imagining the mask $h[i, j]$ being shifted across the input image $f[n, m]$ to different offsets; then the superimposed values at this offset are multiplied together, and the products added. The resulting sum at this offset is the value of the output image $g[n, m] = (f * h)[n, m]$ at the pixel location corresponding to the center of the mask. This is shown in Figure 4.1.

The classical edge detectors, attributed to Prewitt, Sobel and Kirsch, are examples of 2D convolution filters defined by [4.3]. The simplest of the three is the Prewitt filter, which is defined by a real-valued convolution mask over a 3×3 neighborhood as:

$$h = \begin{bmatrix} 1 & 1 & 1 \\ 0 & 0 & 0 \\ -1 & -1 & -1 \end{bmatrix} \qquad [4.4]$$

This mask detects horizontal edges; a transposed version of this mask can be used to detect vertical edges. Both filters can be combined to yield an edge strength output image. This can be done by applying the second filter to the output image of the first, or a combined filter can be created by convolving one mask against the other. Likewise, the classic Sobel filter has the mask values:

$$h = \begin{bmatrix} 1 & 2 & 1 \\ 0 & 0 & 0 \\ -1 & -2 & -1 \end{bmatrix} \quad [4.5]$$

for detecting horizontal edges. Finally, the classic Kirsch filter has the mask values:

$$h = \begin{bmatrix} -3 & 5 & 5 \\ -3 & 0 & 5 \\ -3 & -3 & -3 \end{bmatrix} \quad [4.6]$$

and seven other similar masks, each sensitive to vertical or horizontal edges, or edges inclined at $45°$ to the vertical. Each of these edge detection filters is a high-pass filter, meaning that low-frequency components of the input image are attenuated and high-frequency components accentuated. This can be seen by computing the complex Fourier transforms of their masks, $h[n, m]$, and examining the resulting spectral coefficients, $H[u, v]$, at specific frequencies $(\pm u, \pm v)$.

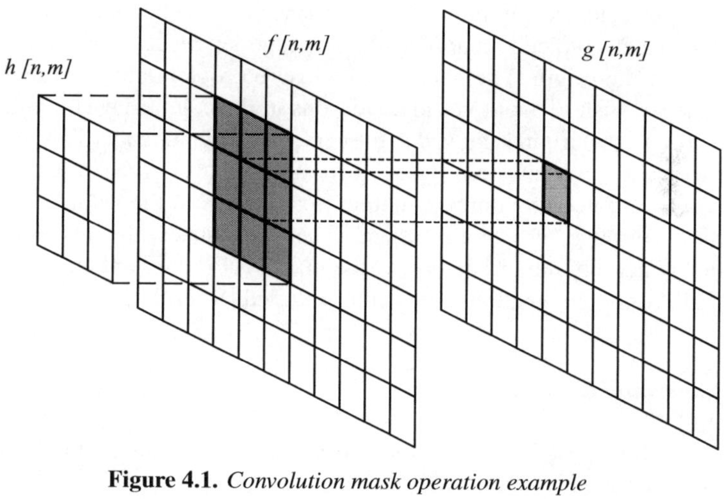

Figure 4.1. *Convolution mask operation example*

The extension of these ideas to color images *could* be achieved by repeated application to each color component of the image, as if each component were a grayscale image. This is, in fact, what is commonly done. But this implies that such marginal, component-wise filtering is sufficient under all desired color image filtering operations. It has been shown that this assumption is false, and the rest of this section shows why.

4.1.2. *Color images as quaternion arrays*

Color image pixels typically have three components, as such they can be represented in quaternion form using pure quaternions[1]. For example, a pixel at image coordinates (n, m) in an RGB color space image can be represented as $f[n, m] = r[n, m]\boldsymbol{i} + g[n, m]\boldsymbol{j} + b[n, m]\boldsymbol{k}$, where $r[n, m]$ is the red component and $g[n, m]$ and $b[n, m]$ are the green and blue components of the pixel, respectively. The same approach may be used with other color space models, for example, a luminance-chrominance color space, such as YC_bC_r [PAL 98]. Using the 3D pure imaginary quaternion is neither coincidental nor arbitrary. As we saw in Chapter 2, a full quaternion may be interpreted as the ratio of two vectors, the quantity that multiplies one vector to give another. This geometric interpretation is exploited in the design of color filters.

In classical image filtering, we typically have to handle the problem of color space *closure*. Image pixel values have a finite range, typically positive and scaled to the interval $[0, 1]$. If a filtering operation results in a pixel value outside this range, then it must be brought back into the range, ideally without distortion artifacts. One says the range of legitimate pixel values is not closed with respect to the filtering operation. For grayscale images, closure is typically forced by clipping the pixel value at the range boundaries. For color image filters, the color space is a 3D bounded volume, e.g., the RGB color space is typically mapped to a unit edge-length cube with one corner (the black color) at the origin. The color space closure problem becomes more difficult, especially if we wish to avoid visual distortions in the color image output. To reduce these distortions, the origin is moved to the center of the color cube. This means that all pixel values are now vectors pointing away from mid-gray, the center of the cube. Now closure can be enforced by clipping output pixel values to the value where the pixel vector passes through the cube's surface. This three-space clipping eliminates hue distortions by holding the orientation of the pixel vector constant during the clipping operation, i.e. only the length of the vector is altered [SAN 04].

4.1.3. *Quaternion convolution*

There are three possible quaternion convolution definitions available: left-, right- and bi-convolution defined, respectively, as:

$$(h_L \circ f)[n, m] = \sum_{i=0}^{N-1} \sum_{j=0}^{M-1} h_L[i, j]\, f[n - i, m - j] \qquad [4.7]$$

1 Some color spaces are four-dimensional but because of the tri-stimulus response of the human visual system most are three-dimensional.

$$(f \circ h_R)[n,m] = \sum_{i=0}^{N-1} \sum_{j=0}^{M-1} f[n-i, m-j]\, h_R[i,j] \qquad [4.8]$$

and

$$(h_L \prec f \succ h_R)[n,m] = \sum_{i=0}^{N-1} \sum_{j=0}^{M-1} h_L[i,j]\, f[n-i, m-j]\, h_R[i,j] \qquad [4.9]$$

where $f[n,m]$ is the input image and $h_L[n,m]$ and $h_R[n,m]$ are the convolution masks. The bi-convolution operation is shown in Figure 4.2 where the filter masks sandwich the input image.

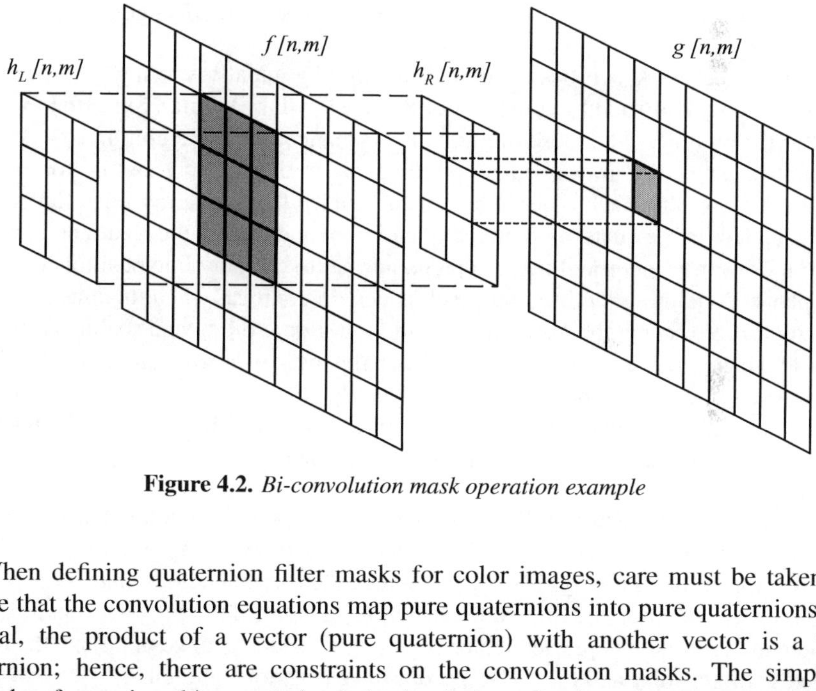

Figure 4.2. *Bi-convolution mask operation example*

When defining quaternion filter masks for color images, care must be taken to ensure that the convolution equations map pure quaternions into pure quaternions. In general, the product of a vector (pure quaternion) with another vector is a full quaternion; hence, there are constraints on the convolution masks. The simplest example of ensuring this constraint is in the design of color edge detection filters. This was first done in [SAN 98c] with a generalization of the Prewitt filter shown in [4.4]. To this design was added two additional filters, generalizations of the Sobel and Kirsch edge detection filters [SAN 00], all based on vector rotation operations. These filters are a generalization of the classical grayscale edge detection filters of Prewitt, Sobel and Kirsch, as described in section 4.1.1. They employ a vector rotation in color space about the gray line of the RGB color space on which all achromatic pixel values lie; hence, they require the use of the bi-convolution equation [4.9].

The Prewitt color filter is defined by two quaternion-valued bi-convolution masks over a 3×3 neighborhood as

$$\alpha \begin{bmatrix} 1 & 1 & 1 \\ 0 & 0 & 0 \\ q & q & q \end{bmatrix} \begin{bmatrix} & & \\ & f & \\ & & \end{bmatrix} \begin{bmatrix} 1 & 1 & 1 \\ 0 & 0 & 0 \\ \overline{q} & \overline{q} & \overline{q} \end{bmatrix} \qquad [4.10]$$

where $q = \exp(\boldsymbol{\mu}\pi/2)$ and $\alpha = 1/6$ is a scale factor to account for the addition of the six non-zero pixel values, and f denotes the input image. This expression is not a matrix equation but a shorthand notation for the situation shown in Figure 4.2. When $\boldsymbol{\mu}$ points along the gray line, the rotation operator $\mathcal{R}_q[\{f\}] = q\ \{f\}\ \overline{q}$ rotates a pixel value $\{f\}$ through π radians in the plane normal to the luminance axis or gray line (see [2.3]). Like the classical Prewitt filter, these masks detect horizontal edges; a transposed version of both masks can be used to detect vertical edges.

Where the mask overlies an image region of gradually varying color, the pixel values along the top and bottom rows of the mask will be similar. Since the top row is multiplied by unity (and therefore remains unchanged) and the bottom row is rotated through π about the gray line, the pixel values in the top row and bottom row are positioned opposite each other at an equal distance from the gray line after rotation. When added (using quaternion or vector addition) and scaled, the resultant pixel value lies on or near to the gray line, in a plane normal to the gray line positioned between the planes containing the original pixel values. In contrast, when the mask overlies a horizontal color edge, the result after rotation, vector addition and scaling does not lie near to the gray line and exhibits color. When the mask overlies a horizontal luminance edge where the pixels above and below the edge have similar or identical hues, the chrominance components of the pixels will almost cancel. The cancellation is less perfect where there is a large difference in luminance.

Figure 4.3, provided in full color in the color section within this book, shows the result of applying the Prewitt filter to an image. The filtered image is a full color image, despite appearing as a grayscale image. Where the original color image had gradually varying areas of color, the filtered image has been reduced to gray (achromatic) pixels. Where the original image had a color edge (i.e. a sudden change of color), the filtered image exhibits color and the "edge" pixels have a vector "strength".

The Sobel and Kirsch filters, ([4.5] and [4.6], respectively), are likewise generalized with the expressions:

$$\frac{1}{8} \begin{bmatrix} 1 & \sqrt{2} & 1 \\ 0 & 0 & 0 \\ q & \sqrt{2}q & q \end{bmatrix} \begin{bmatrix} & & \\ & f & \\ & & \end{bmatrix} \begin{bmatrix} 1 & \sqrt{2} & 1 \\ 0 & 0 & 0 \\ \overline{q} & \sqrt{2}\overline{q} & \overline{q} \end{bmatrix} \qquad [4.11]$$

and

$$\frac{1}{30}\begin{bmatrix} \sqrt{3}q & \sqrt{5} & \sqrt{5} \\ \sqrt{3}q & 0 & \sqrt{5} \\ \sqrt{3}q & \sqrt{3}q & \sqrt{3}q \end{bmatrix}\begin{bmatrix} \\ f \\ \\ \end{bmatrix}\begin{bmatrix} \sqrt{3}\bar{q} & \sqrt{5} & \sqrt{5} \\ \sqrt{3}\bar{q} & 0 & \sqrt{5} \\ \sqrt{3}\bar{q} & \sqrt{3}\bar{q} & \sqrt{3}\bar{q} \end{bmatrix} \qquad [4.12]$$

where q is as in the Prewitt filter.

Figure 4.3. *Filtered output (right) of color image (left) using Prewitt edge detector (see color section within this book for a full color version of this figure)*

Finally, it is worth noting that the color of the edges in the *output* image is related to the colors involved in the edge transitions of the input image. That is, the colors of regions on either side of the edge determine the color of the final output edge value. Also, reversing the transition, e.g., red to blue *versus* blue to red, will complement the output edge color.

4.1.4. *Quaternion image spectrum*

Using the left-sided QFT of [3.41], with transform axis chosen to be $\boldsymbol{\mu} = (\mathbf{i}+\mathbf{j}+\mathbf{k})/\sqrt{3}$ (which points along the gray line in color space), we obtain the quaternion image spectrum. This quaternion spectrum is an array of quaternion values, with one real and three imaginary parts. We customarily treat this spectrum as a quaternion image – that is, we regard each spectral coefficient as a quaternion pixel. It is usual for all four parts of any given spectral coefficient to be non-zero. A similar statement applies to the complex spectra of grayscale images, where the coefficients usually have non-zero real and imaginary parts, even when the image pixels are purely real.

Complex spectra are interpreted and visualized by creating images based on writing the spectrum in polar form: one image for modulus and the other for phase.

In a similar manner, a meaningful visualization of the quaternion spectrum can be based on quaternion polar form (see [1.47]). However, quaternions in polar form have a third quantity besides modulus and phase: i.e. the axis. This has no counterpart in the complex case because the imaginary operator $I = \sqrt{-1}$ in the complex case is constant. In the quaternion polar form, this imaginary operator $\boldsymbol{\mu} = \sqrt{-1}$ may be any pure unit quaternion.

The modulus is visualized as a grayscale image, appropriately scaled using, for instance, a logarithmic scaling such as

$$l(q) = \frac{\log(1 + |q|)}{\log(1 + K)}$$

where K is the largest modulus in the image. The phase can be visualized by using a color representation based on the IHS color space. We create an RGB value from an IHS value, where H is the phase, I is the mid-range and S is unity. This means that a phase of zero is represented by a red hue, a phase of $\pi/2$ is represented by a green hue and a phase of π is represented by a green-cyan hue. This leaves us with the axis to represent. Because $\boldsymbol{\mu}$ is a pure quaternion, it has three components. This suggests a representation using an RGB image. The standard RBG color space is restricted to one octant of 3-space, as each color component must be non-negative. The values of $\boldsymbol{\mu}$, however, may be anywhere on a unit sphere centered at the origin. By scaling this sphere by $\frac{1}{2}$ and placing it at the center of a unit edge-length color cube, we can represent all possible values of $\boldsymbol{\mu}$. To further enhance the image, the $\boldsymbol{\mu}$ value is scaled until it extends to the surface of the color cube, so that the resulting image is fully saturated. This is done by scaling $\boldsymbol{\mu}$ so that the largest component has unit value. Figure 4.4, provided in full color in the color section within this book, shows the quaternion spectrum of an image, visualized as described in the previous text, with the constant term in the center of the image, as is conventional. Positive frequencies are, therefore, in the upper right quadrant.

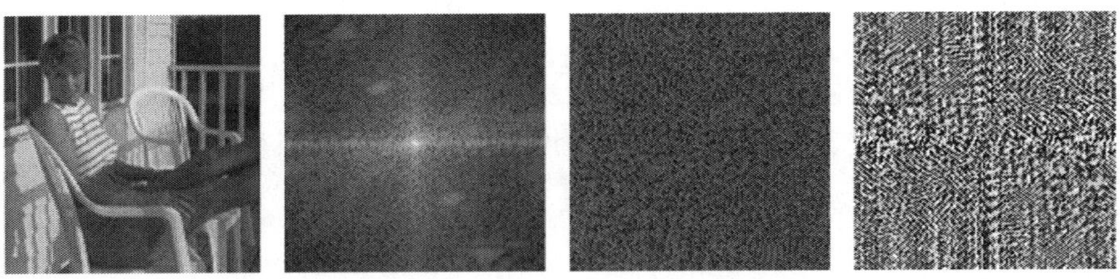

Figure 4.4. *Quaternion spectrum of an image. Left to right: original image, spectral modulus, phase and axis (see color section within this book for a full color version of this figure)*

Given a quaternion spectrum image, $F[u,v]$, its pixel-by-pixel symplectic decomposition with respect to $\mathbb{C}_{\boldsymbol{\mu}}$ is given as

$$F[u,v] = F_1[u,v] + F_2[u,v]\boldsymbol{\mu}_2$$

where each part F_1 and F_2 are in $\mathbb{C}_{\boldsymbol{\mu}}^{N\times M}$. The orthogonal vector $\boldsymbol{\mu}_2$ can be placed anywhere in the chromatic plane, but in the pictured example, it is parallel to the red line. If one takes the inverse transform of each component: $f_1 = \mathcal{F}^{-1}\{F_1\}$ and $f_2 = \mathcal{F}^{-1}\{F_2\}$, then it happens that the simplex part f_1 contains the luminance information and the perplex part f_2 contains the chrominance information. This is not a coincidence, as $\boldsymbol{\mu}$ is parallel to luminance and everything perpendicular to $\boldsymbol{\mu}$ is in the chrominance plane. The simplex part will be purely imaginary, luminance requiring only one real number. The perplex part will be complex, chrominance requiring two real numbers. Figure 4.5, provided in full color in the color section within this book, shows the previous image decomposed in this manner. Both parts are still full color images, not grayscale images. The original image can be recovered by adding the simplex and perplex parts.

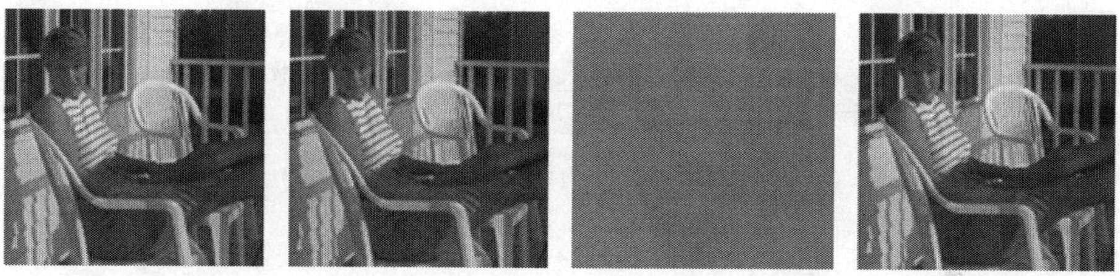

Figure 4.5. *Symplectic decomposition of color image into simplex and perplex parts of spectra. Left to right: original image, simplex part, perplex part and direct sum of both parts (see color section within this book for a full color version of this figure)*

The arrangement of coefficients in the quaternion spectrum follows the same layout as that in a complex Fourier transform spectrum; the spectrum is thus divided into quadrants as shown in Figure 4.6. Any given pair of horizontal and vertical frequencies is represented in the spatial frequency domain by two quaternion values, the pair being taken from diagonal quadrants. Exceptions are the Nyquist frequency pair, and the DC term, and any pair of frequencies including a DC or Nyquist term. (The coefficients in the two quadrants correspond to combinations of positive and negative horizontal and vertical spatial frequencies.) If the image is to be a pure quaternion image, which will be the case if color images are represented by the color components in the three components of the vector part of a quaternion pixel, the scalar parts of the two frequency components must cancel out. Certain symmetry conditions on the image (analogous to conjugate symmetries in complex images) will ensure this; in general, an image with no symmetries whatever (e.g. a random image) must have a spectral domain representation that satisfies this constraint; otherwise the inverse QFT of the spectral coefficients would not reconstruct the original image (the scalar part would be non-zero).

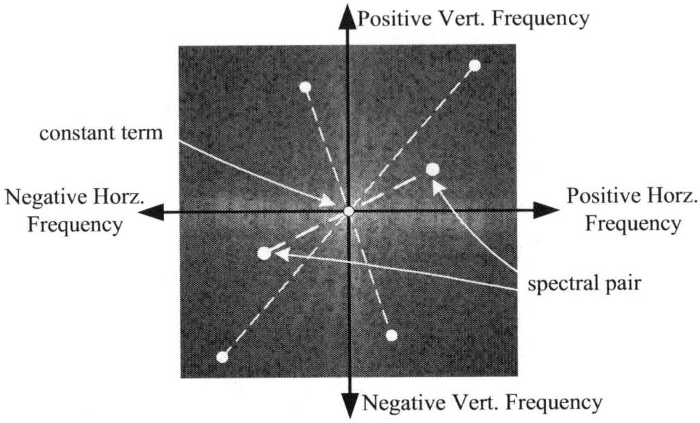

Figure 4.6. *Spectral coefficient layout*

Consider the inverse transform that reconstructs an image from a pair of spectral coefficients and the basis functions:

$$f[n,m] = \sum_{v=0}^{M-1}\sum_{u=0}^{N-1} e^{\boldsymbol{\mu}2\pi\left(\frac{mu}{M}+\frac{nu}{N}\right)}\, F[V,U]$$

and extract two terms from the summation, which make up the component of the image at specific horizontal and vertical spatial frequencies. Let the arbitrary frequencies, denoted by their index values, be V and U (where $0 < V < M/2$ and $0 < U < N-1$). From the symmetry of the coefficients, those at $M-V$ and $N-U$ also represent the same frequencies; so for a given choice of V and U, we have, after some simplification:

$$\begin{aligned} f'[n,m] &= e^{\boldsymbol{\mu}2\pi\left(\frac{mu}{M}+\frac{nu}{N}\right)}\, F[V,U] + e^{\boldsymbol{\mu}2\pi\left(\frac{-mu}{M}+\frac{-nu}{N}\right)}\, F[M-V,N-U] \\ &= e^{\boldsymbol{\mu}2\pi(+\alpha+\beta)}\, F[V,U] + e^{\boldsymbol{\mu}2\pi(-\alpha-\beta)}\, F[M-V,N-U] \end{aligned}$$

Any coefficient of the Fourier spectrum may be separated into simplex and perplex parts with respect to $\boldsymbol{\mu}$:

$$F[V,U] = F_s[V,U] + F_p[V,U]\boldsymbol{\mu}_2, \quad \text{where } F_s, F_p \in \mathbb{C}_{\boldsymbol{\mu}}.$$

The simplex and perplex parts of the coefficient multiply the exponential basis function of the Fourier transform to produce a cosinusoidal image component with horizontal and vertical frequencies determined by the indices of the coefficient, and with amplitude, phase and orientation in color space determined by the value of the coefficient.

$$\begin{aligned} f'[n,m] &= e^{\boldsymbol{\mu}2\pi(+\alpha+\beta)}\{F_s[V,U]+F_p[V,U]\boldsymbol{\mu}_2\} \\ &\quad + e^{\boldsymbol{\mu}2\pi(-\alpha-\beta)}\{F_s[M-V,N-U]+F_p[M-V,N-U]\boldsymbol{\mu}_2\} \\ &= e^{\boldsymbol{\mu}2\pi(+\alpha+\beta)}F_s[V,U]+e^{\boldsymbol{\mu}2\pi(-\alpha-\beta)}F_s[M-V,N-U] \\ &\quad + e^{\boldsymbol{\mu}2\pi(+\alpha+\beta)}F_p[V,U]\boldsymbol{\mu}_2+e^{\boldsymbol{\mu}2\pi(-\alpha-\beta)}F_p[M-V,N-U]\boldsymbol{\mu}_2 \end{aligned}$$

Figure 4.7. *Spectral coefficient images for the constant term and first three harmonics (horizontally and vertically) of color image. The DC term is the center tile on bottom row with positive horizontal frequencies to the right and negative horizontal frequencies on the left. The bottom row corresponds to the DC vertical term with each previous row increasing in vertical frequency – the left and right halves of the last row are mirrors of each other as they contain the same spectral coefficient pairs (see color section within this book for a full color version of this figure)*

Since the first two terms in the last equality are isomorphic to the complex numbers (being in $\mathbb{C}_{\boldsymbol{\mu}}$), we see that the coefficients $F[V,U]$ and $F[M-V,N-U]$ behave analogously to the classical complex case. That is, these coefficients scale the magnitude and alter the phase of the Fourier exponentials that combine to form a cosinusoidal function. The same is true for the remaining terms in the last equality, after factoring out the $\boldsymbol{\mu}_2$ term from both. This is shown in Figure 4.7, which is provided in full color in the color section within this book. When these two cosinusoidal functions are added, they describe an orbit in the color space centered at mid-gray. A demonstration of the color orbit interpretation of the spectral components can be made possible through the use of a color-cube scatter plot. If each pixel value of an image is plotted in color space, the resulting scatter plot shows the distribution of the image's color contents in color space. The first row of Figure 4.8, provided in full color in the color section within this book, shows an image and its color-cube scatter plot. If the original image is quaternion Fourier transformed, a single harmonic (in one direction) extracted and inverse transformed, the resulting image will show a "rainbow grating" pattern due to harmonic oscillations in luminance and chrominance. The scatter plot of this rainbow grating image should draw an elliptic orbit about the center of the cube. The second through sixth rows of Figure 4.8 show that, indeed, each harmonic describes an elliptical orbit in the color space, which results in a rainbow grating at that harmonic. The rightmost column shows the resulting image built up by progressively adding each new harmonic (horizontal and vertical) to the previous harmonics.

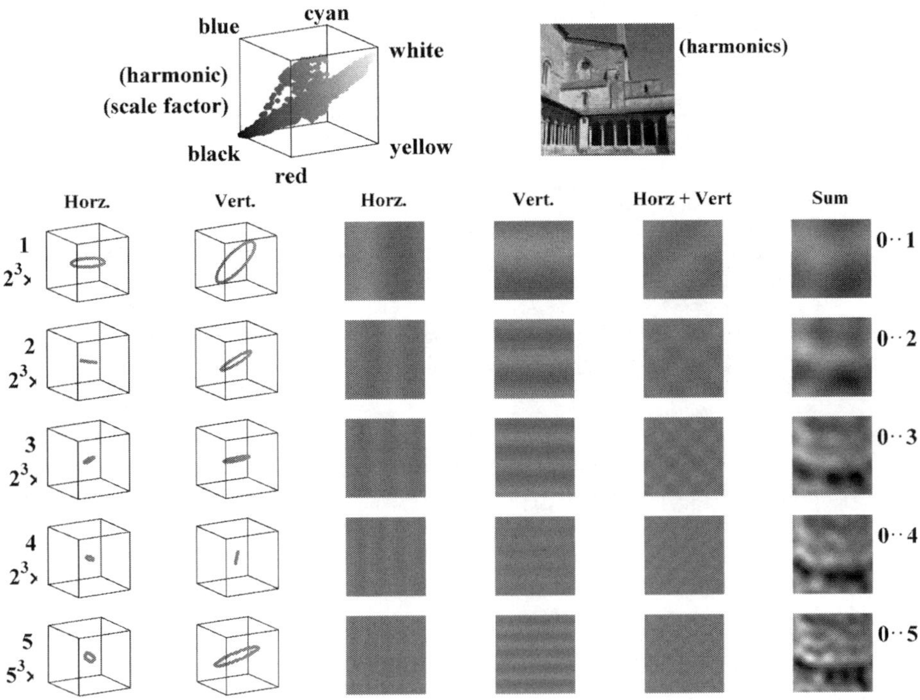

Figure 4.8. *Harmonic reconstruction of color image. The legend at the top shows a color-cube scatter plot for the original image. The lower grid shows plots of the color orbits for the first five horizontal and vertical harmonics taken from the QFT spectrum of the image, along with their corresponding images. The final, rightmost column shows the image resulting from the sum of the harmonics, starting from the constant (zero frequency) term to the corresponding harmonic (see color section within this book for a full color version of this figure)*

Figure 4.9, provided in full color in the color section within this book, shows results of filtering the image directly in the quaternion spectral domain. The first column shows the original image and its spectral modulus. The second column shows the results of low-pass filtering the image by zeroing the 12th and higher spectral harmonics. As expected, the resulting image is blurred consistent with this operation. The third column repeats the process in bandpass fashion, clearing the constant and first two harmonics as well as the 26th and higher harmonics. As expected, only moderate-sized regions with uniform luminance or chrominance can be detected in the final image. Finally, the last column shows high-pass filtering by zeroing the first five harmonics. Again, as expected, the image contains content only where there is rapid luminance or chrominance variation.

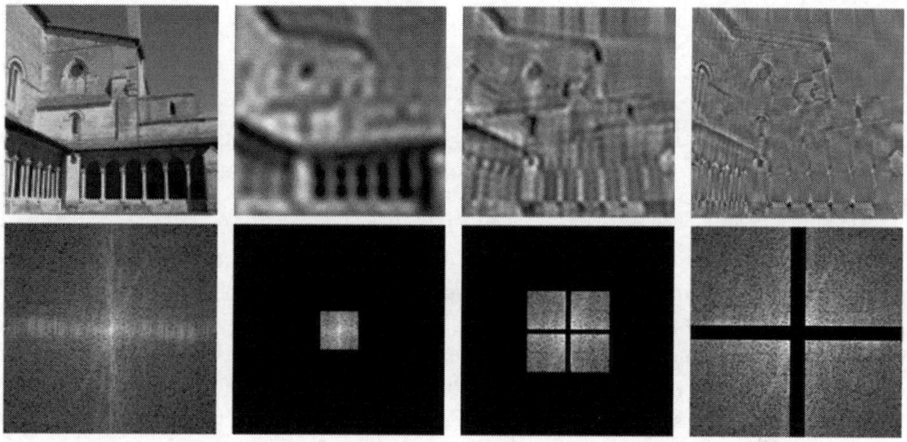

Figure 4.9. *Direct filtering in frequency domain of color image. Top row, left to right: original image, low-pass, band-pass and high-pass filtered. The bottom row shows the corresponding spectral magnitude (see color section within this book for a full color version of this figure)*

4.2. Generalized correlation

Correlation is concerned with detecting and locating *similarity* in signals or images. Given a known *feature* such as a pulse of a particular shape and duration, or, in an image, an object seen from a known viewpoint, correlation can locate the feature, if present, within a longer, noisy, signal; or within a noisy or degraded image. For example, it may be desired to locate a *target*[2] object within a defocused or blurred image; or in the context of color image processing, an image taken under

2 The use of the word "target" here is not meant to imply a military application, although of course, there are military applications for object location within images.

lighting conditions such that the colors of the target object are different from a reference image. More precisely, correlation will reveal the *most likely* location, or multiple locations, of the feature. A second use for correlation is to compute the temporal offset between two similar signals or the spatial offset between two images in order to bring one into *registration* with the other by applying a shift to one signal or image equal to the offset between them. As with any other fundamental technique, correlation can be used as a step within a more complex algorithm.

A closely related concept is *phase correlation* [KUG 75] in which the normalized *cross-power spectrum* is computed, from which the inverse Fourier transform yields a correlation peak at an offset representing the best correlation between the two signals or images. Phase correlation works particularly well with images, where sharp edges are frequently present (these have frequency content that can be exploited well by the phase correlation technique).

Correlation was generalized to quaternion signals and images in 2003 [MOX 03], and we consider it in detail in section 4.2.2. Phase correlation was successfully generalized to color images using QFTs in 2001 [SAN 01], and we consider it in detail in section 4.2.3.

In the next section, we discuss the classical cases of correlation and phase correlation, before introducing some quaternion equivalents. The classical case is normally presented in the literature for real-valued signals, but in fact it can easily be generalized to complex-valued signals, which is a more useful starting point for the quaternion case in this book. The principal difference concerns the relationship between time (or spatial) domain reversal (of a signal or image), and the corresponding effect on the frequency-domain representation. Time reversal of a signal corresponds, in general, to interchange of the positive and negative frequencies in the spectrum of the signal. However, the Fourier spectrum of a real signal has conjugate symmetry, that is the positive and negative frequency coefficients are conjugates of each other, and therefore the effect of interchanging positive and negative frequencies can be obtained more easily (without moving data) by conjugating the Fourier coefficients, and it is the latter operation that is mentioned in most texts.

Correlation can be defined (and computed numerically) in both the time and frequency domains, but phase correlation has to be computed in the frequency domain, because it requires the phase of the Fourier coefficients[3]. There is an equivalence between the two domains. In each domain, there is a formula – in the time domain, this requires an integral (or summation in the discrete case), whereas in

3 There is a normalization involved using the modulus of the products of the Fourier coefficients, and this operation is not linear. It does not therefore have a straightforward equivalent in the time domain.

Figure 4.3. *Filtered output (right) of color image (left) using Prewitt edge detector*

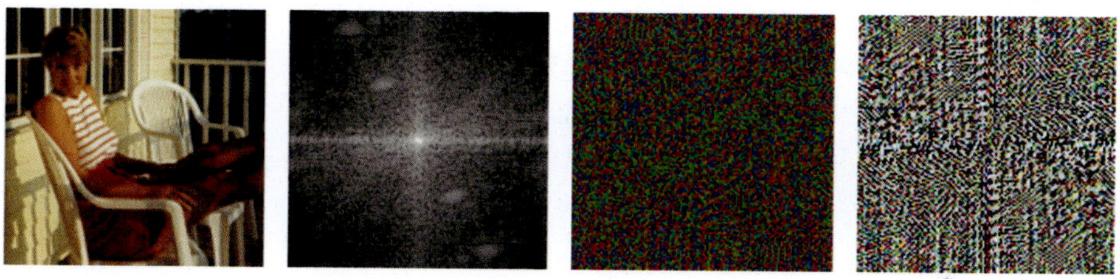

Figure 4.4. *Quaternion spectrum of an image. Left to right: original image, spectral modulus, phase and axis*

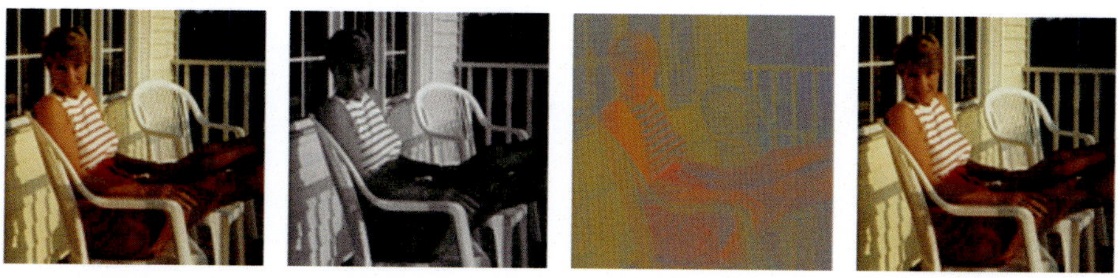

Figure 4.5. *Symplectic decomposition of color image into simplex and perplex parts of spectra. Left to right: original image, simplex part, perplex part and direct sum of both parts*

Figure 4.7. *Spectral coefficient images for the constant term and first three harmonics (horizontally and vertically) of color image. The DC term is the centre tile on bottom row with positive horizontal frequencies to the right and negative horizontal frequencies on the left. The bottom row corresponds to the DC vertical term with each previous row increasing in vertical frequency. (The left and right halves of the last row are mirrors of each other as they contain the same spectral coefficient pairs.)*

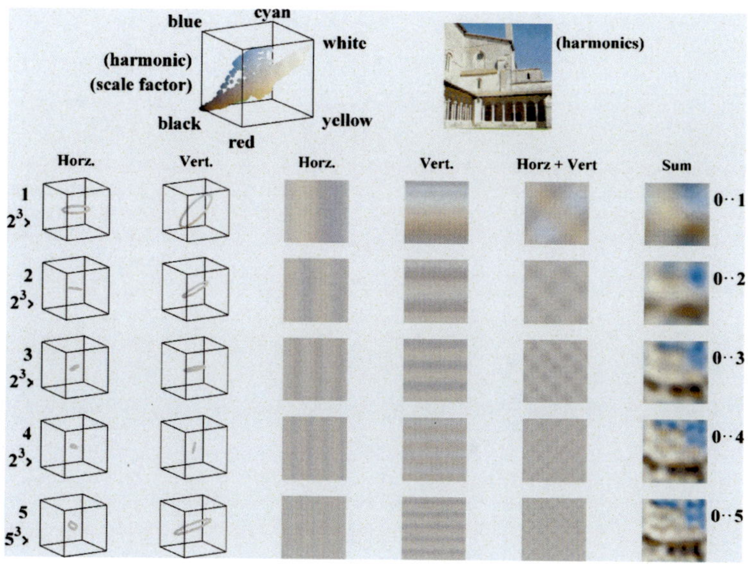

Figure 4.8. *Harmonic reconstruction of color image. The legend at top shows a color-cube scatter plot for the original image. The lower grid shows plots of the color orbits for the first five horizontal and vertical harmonics taken from the QFT spectrum of the image, along with their corresponding images. The final, right-most column shows the image resulting from the sum of the harmonics, starting from the constant (zero frequency) term to the corresponding harmonic*

Figure 4.9. *Direct filtering in frequency domain of color image. Top row, left to right: original image, low-pass, band-pass and high-pass filtered. The bottom row shows the corresponding spectral magnitude*

Figure 4.12. *Two color images with identical backgrounds but a foreground object (the car) in different positions with different colors. (Images by and reproduced with the permission of Dr A.L. Thornton [THO 98])*

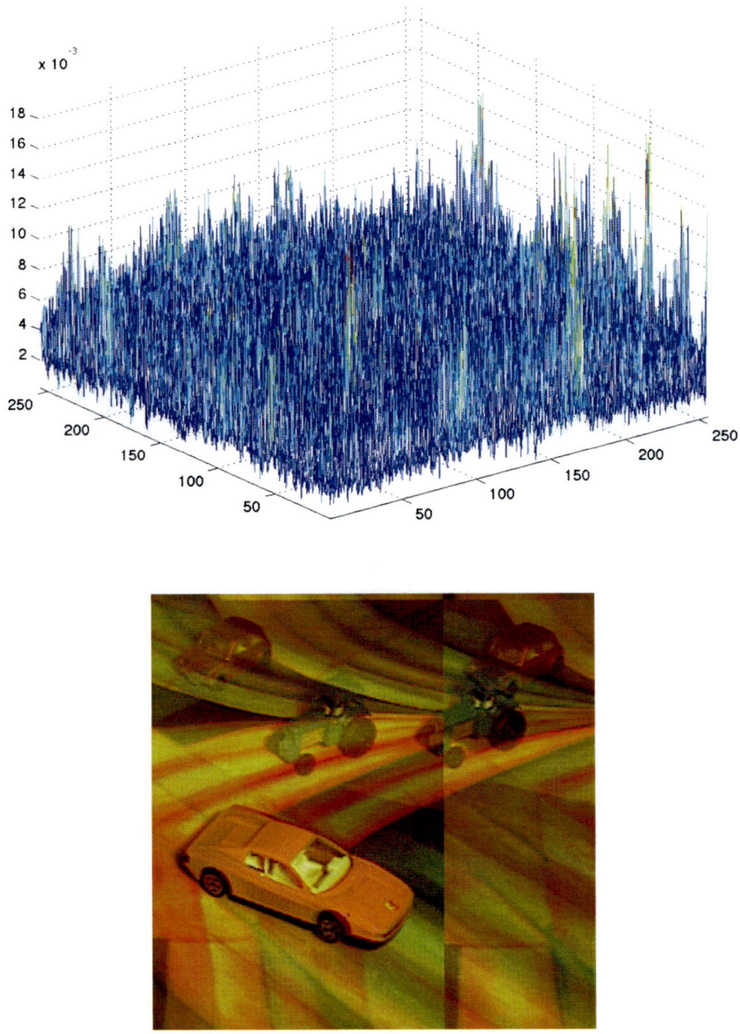

Figure 4.13. *Phase correlation results using the color images in Figure 4.12. Upper: the phase correlation surface. Lower: the two images superimposed after shifting one of them by the coordinate offset found from the maximum peak in the phase correlation surface, showing the two cars correctly superimposed*

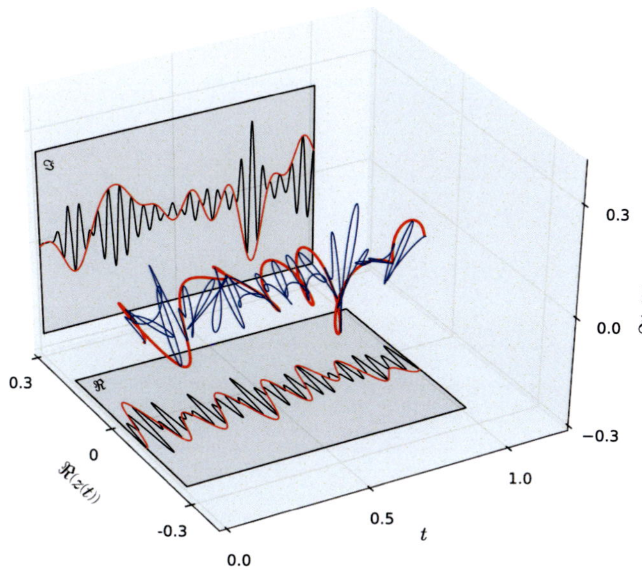

Figure 4.19. *Orthocomplex modulated complex signal with constant frequency $z(t)$. 3D representation of the complex signal $z(t)$ (blue solid line) and its complex instantaneous amplitude $\rho_z(t)$ (red solid line) obtained from the polar CD form of its quaternion extension $q_z(t)$. The projections in $\Re$ and $\Im$ planes (black solid lines) correspond to the signals presented in Figure 4.20*

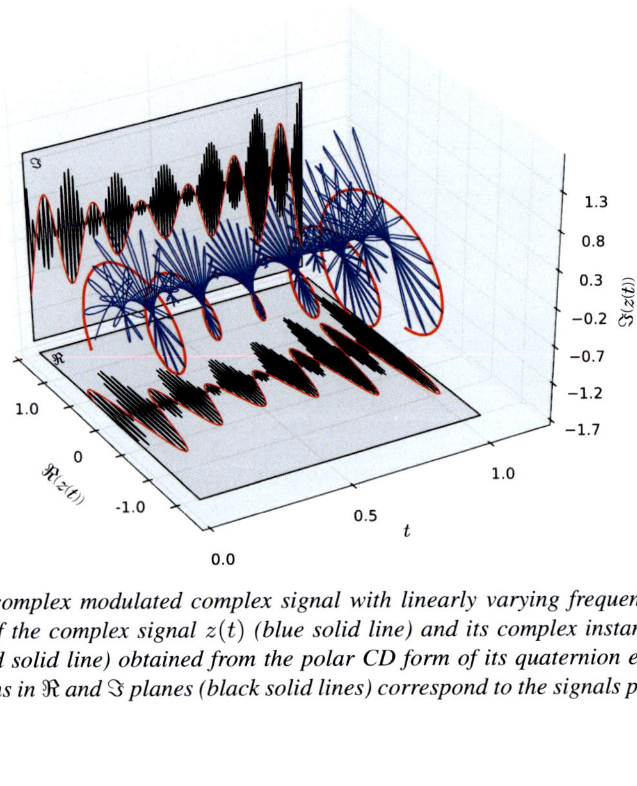

Figure 4.21. *Orthocomplex modulated complex signal with linearly varying frequency $z(t)$. 3D representation of the complex signal $z(t)$ (blue solid line) and its complex instantaneous amplitude $\rho_z(t)$ (red solid line) obtained from the polar CD form of its quaternion extension $q_z(t)$. The projections in $\Re$ and $\Im$ planes (black solid lines) correspond to the signals presented in Figure 4.22*

the frequency domain it is a point-wise product between Fourier coefficients. The difference is computationally significant (the Fourier domain approach is much faster for practical signals and images). However, it is necessary to know the frequency domain formula, and this depends on the QFT(s) used. In the quaternion cases presented after the next section, it will be shown that the formula is not a trivial generalization of the classical case. Formulas of this type are not easily discovered, as was noted by J.L. Synge in 1972:

> It is typical of quaternion formulae that, though they be difficult to find, once found they are immediately verifiable. [SYN 72]

One of the ways that these formulas are verifiable is to implement them, of course, and check that the implementation behaves as expected, but a very useful fact to remember is that the formula must reduce to the classical case when the quaternion variables within it are constrained to a subspace that is isomorphic to the complex numbers, for example, when all the quaternions within the formula have the same axis (direction of the vector part).

4.2.1. *Classical correlation and phase correlation*

Cross-correlation[4] can be defined directly in the time domain (without Fourier transforms) using the following formula [WEI 14a]:

$$c(t) = \int_{-\infty}^{\infty} f^{\star}(\tau) g(t+\tau)\,\mathrm{d}\tau \qquad [4.13]$$

The conjugate in the formula is vital if the signals are complex; otherwise, the product of the imaginary parts of two similar signals could cancel with the products of the real parts. Conjugating one of the signals negates one of the imaginary parts and thus cancels out the -1 arising from the product of the two imaginary roots of -1. Of course, if the signals are real, the conjugate has no effect. Note that the formula is equivalent to convolution (compare [4.2]) with one signal time-reversed and conjugated [WEI 14a]. Note that one of the signals must be shifted through all possible shifts relative to the other signal. In the discrete case (with signals of finite length), the formula becomes:

$$c[n] = \sum_{m=0}^{M-1} f^{\star}[m] g[n+m] \qquad [4.14]$$

4 If the mean, or zero-frequency, component of the signals is subtracted out (i.e. f and g in the formula have zero means), then the result is known as the *cross-covariance*.

assuming the signals are of the same length, M samples. For any reasonable length of signal or size of image, a much faster calculation is possible using Fourier transforms. In order to do this, it is necessary to know the operational formula that expresses the equivalent of [4.14] in the frequency domain. In the classical case, this is:

$$c[n] = \mathcal{F}^{-1}\left\{\mathcal{F}^{\star}\{f\}\,\mathcal{F}\{g\}\right\} \tag{4.15}$$

Note that the product of the two Fourier spectra is computed element-wise (sometimes known as point-wise), that is, each coefficient in the spectrum is multiplied by the corresponding coefficient in the other spectrum (and one of the two is also conjugated). A subtlety about the use of discrete Fourier transforms is that they assume that the signal is periodically repeated over all time (in image processing this is often referred to as *tiling* – the image is assumed to repeat over an infinite plane both horizontally and vertically). A consequence is that the formula in the time domain must be modified if it is to yield exactly the same result as the formula based on discrete Fourier transforms, by implementing cyclic wrap around of the shifted signal. This is easily done by modulo arithmetic on the sample index, so that the value $m - n$ in [4.14] is computed modulo M, and hence always has a result falling in the range between 0 and $M - 1$. A more common approach with real signals is to use *zero-padding*.

Figure 4.10 illustrates classical cross-correlation with complex signals by showing the cross-correlation of two noisy signals, each containing a Gaussian-modulated complex exponential pulse. The cross-correlation was computed using [4.15]. The correlation result is complex, with a waveform matching the two signals that were correlated, and a position along the time axis corresponding to the displacement in time between the two original pulses.

Phase correlation is a simple modification of the above ideas, requiring only one extra step. Originally developed for grayscale image processing [KUG 75], the technique works best on signals with edges, rather than smoothly varying signals. (Images tend to have sharp transitions between amplitude levels – at least in a properly focused image, due to the edges of objects.) The technique was shown to work for complex images with pixels representing the chrominance components of a color image [SAN 98b, section 12.2.1]. In the complex case, given two signals or images f and g, the formula required to compute phase correlation is:

$$\mathcal{F}^{-1}\left\{\frac{F^{\star}\,G}{|F^{\star}\,G|}\right\} \tag{4.16}$$

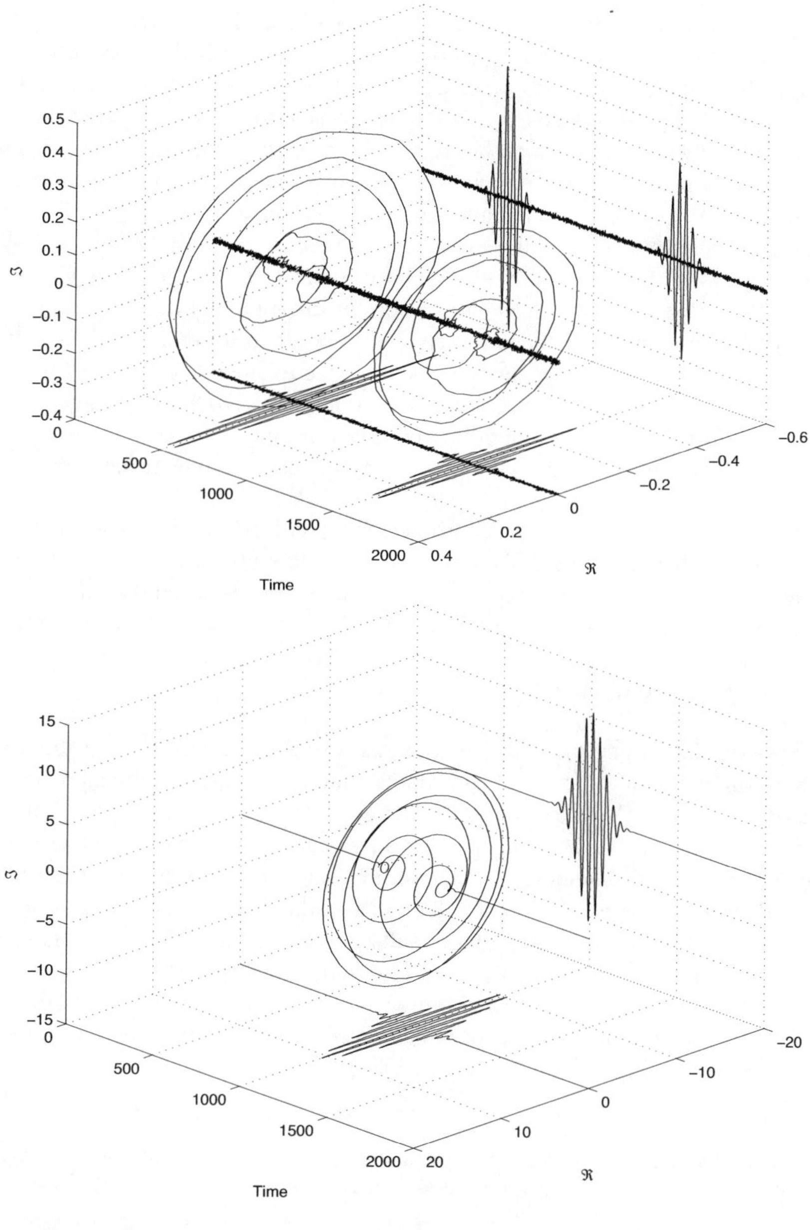

Figure 4.10. *Complex cross-correlation of two Gaussian-modulated complex exponentials. Upper: original signals. Lower: the complex cross-correlation result. In both plots, the real and imaginary parts are projected onto the lower plane and rear plane of the bounding axes*

where $F = \mathcal{F}\{f\}$ and $G = \mathcal{F}\{g\}$. The normalized element-wise or point-wise product of $F^\star$ and G (whose inverse Fourier transform yields the phase correlation) is called the *cross-power spectrum*. Care must be taken with the normalization because a divide-by-zero is possible if any frequency coefficient has zero amplitude (this is unlikely with a real signal, but it can easily happen with simulated signals) [5]. Usually, further processing is applied to the result to extract the location of the phase correlation peak or peaks. The simplest method is to take the modulus of the phase correlation result and look for the maximum sample or pixel value by thresholding. Figure 4.11 shows the result of applying phase correlation to a pair of signals similar to those illustrated in Figure 4.10. An important difference is that the amplitude of the complex exponential has been discretized in order to introduce "edges" into the signal. The phase correlation technique is less suited to smoothly varying signals, but well suited to signals with abrupt amplitude transitions. Notice the sharp peak that results from suppression of modulus information in the cross-power spectrum (essentially the result is based on the phase only, hence the name of the technique).

Note that the location of the peak produced by phase correlation can sometimes be determined to subsample or subpixel accuracy even though the image is sampled on a discrete grid. (As a simple example, consider what happens when the offset between the two signals is half a sample period: the correlation peak will have equal amplitude at two adjacent samples and therefore the location of the peak can be inferred as midway between the two samples).

Cross-correlation and phase correlation, as we have seen from the examples just presented, are defined for both real and complex signals or images. To extend the ideas to vector-valued signals or images (with three or four components per sample), it is obviously possible to compute independent classical correlations on each component (marginal processing). It should be clear without much thought that this is unlikely to work well, although in many cases it will yield a workable result. An obvious difficulty is that three or four separate marginal results will be obtained, and these have to be combined in some way, meaning that an additional algorithmic step is needed, such as a majority vote, or some numerical combination of the marginal results. In this book, we argue that a hypercomplex approach is a much more sensible way to proceed, and correlation is no exception.

5 Obviously, there can also be numerical problems with frequency coefficients that have very small values, and it is wise to apply a threshold criterion to suppress such coefficients.

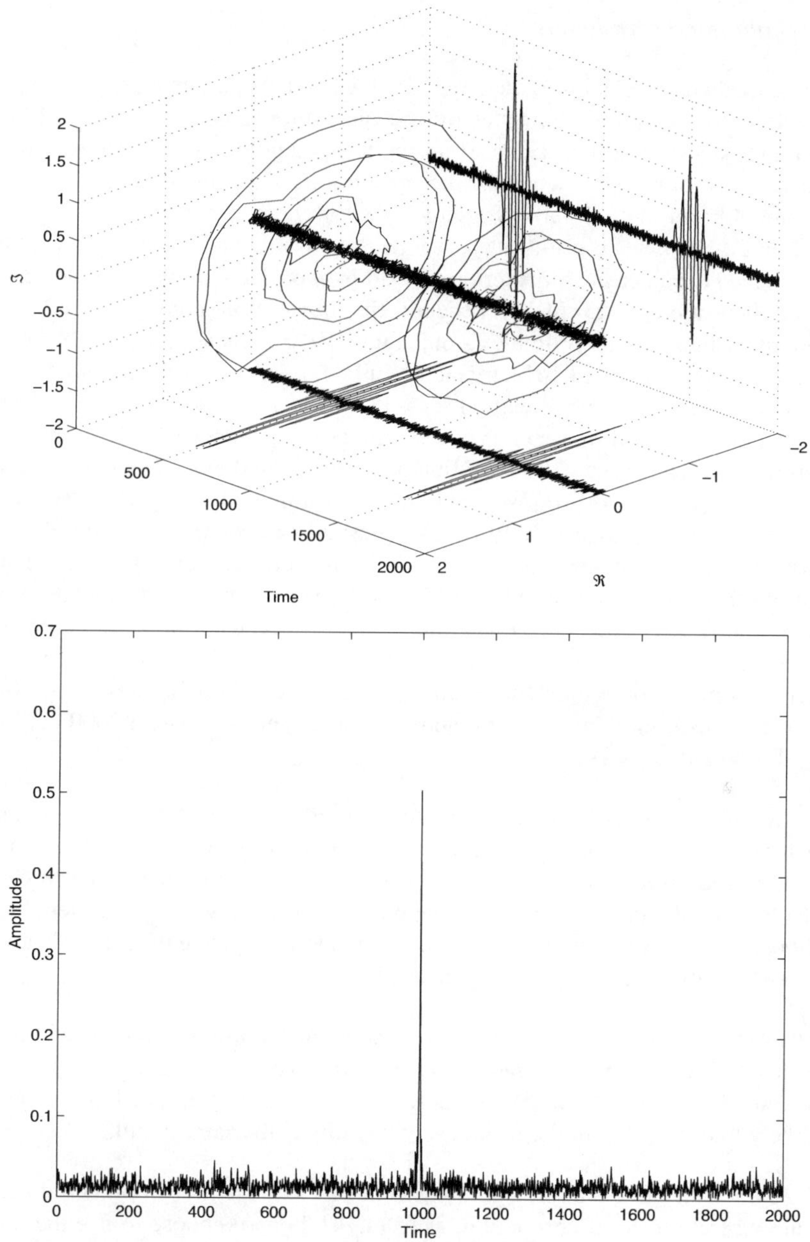

Figure 4.11. *Phase-correlation example. Upper: Gaussian modulated complex exponential pulses with discretized amplitude. Lower: modulus of the phase correlation, showing a sharp peak corresponding to the displacement between the two pulses*

4.2.2. *Quaternion correlation*

To generalize correlation to quaternion-valued signals and images, we use [4.13], and replace the complex conjugate by the quaternion conjugate. That this is the correct approach should be evident from the following: the formula must reduce to the classical complex case when all the quaternion values involved come from the same subfield (i.e. they all have the same *axis*[6] in their vector parts). This is easily seen to be the case, since if this is so, we may replace the axis value by the complex root of -1 and replace each quaternion by an isomorphic complex number, and the formula then reduces to [4.13] because the quaternion conjugate under these conditions reduces to the complex conjugate. We can then immediately compute cross-correlation using [4.14], replacing all the complex operations by their quaternion counterparts. As already discussed, this is a slow method, particularly for images where the run-time depends on the square of the image size, but it provides a very useful reference case for verification of a frequency domain implementation. To work in the frequency domain, we need the quaternion equivalent of [4.15], which is unfortunately not obtained simply by changing all the complex operations to their quaternion counterparts. Furthermore, the frequency domain formula depends on our choice of QFT. Unlike the complex case, where there is only one formulation of the Fourier transform, in the quaternion case, because of non-commutative multiplication, there are many possibilities. Section 3.1.5 explained that a frequency-domain correlation formula may be best derived using a mix of left- and right-sided transforms. This was first done for the quaternion case in 2000 [ELL 00b] using 2D one-sided QFTs.

What follows therefore is an example of the derivation of a frequency-domain correlation formula. A classical complex derivation is given in [WEI 14b], and we follow a similar process here, but in the reverse order, because the example given in [WEI 14b] would require us to choose the Fourier transforms as the first step. Instead, we choose the inverse transform as the first step and manipulate the integrals until we can see which forward transforms are needed.

At various points in this derivation we could make a different choice, and thereby arrive at a different end result. Practically, once we have one formula that is correct and implemented, there is little to gain by deriving others, since the speed of computation is unlikely to differ, and all the formulas will compute the same result.

We start with the quaternion equivalent of [4.13]. This derivation could be done with the discrete form of correlation, as in [4.14], but we choose to use the integral

6 See [1.36] for the definition of the axis of a quaternion.

form in order to make use of the results given in section 3.1.1. Hence, we have the following definition of quaternion cross-correlation:

$$c(t) = \int_{-\infty}^{\infty} \overline{f(\tau)} g(t+\tau)\, \mathrm{d}\tau \qquad [4.17]$$

where c, f and g are now quaternion-valued signals. Take the right-sided QFT of both sides, as in [3.5], but for simplicity, we will omit the scale factor, and we will also leave implicit the transform axis ($+\boldsymbol{\mu}$ unless stated otherwise):

$$\mathcal{F}^R\{c(t)\} = r(\omega) = \int_{-\infty}^{\infty} \left[\int_{-\infty}^{\infty} \overline{f(\tau)} g(t+\tau)\, \mathrm{d}\tau \right] e^{\boldsymbol{\mu}\omega t}\, \mathrm{d}t$$

Since the exponential is constant with respect to the inner integration, and $\mathrm{d}\tau$ is real (and therefore commutes with the exponential), we may re-arrange to obtain:

$$r(\omega) = \int_{-\infty}^{\infty} \int_{-\infty}^{\infty} \overline{f(\tau)} g(t+\tau) e^{\boldsymbol{\mu}\omega t}\, \mathrm{d}\tau\, \mathrm{d}t$$

Changing the order of integration and moving $\overline{f(\tau)}$ outside the new inner integral:

$$r(\omega) = \int_{-\infty}^{\infty} \overline{f(\tau)} \left[\int_{-\infty}^{\infty} g(t+\tau) e^{\boldsymbol{\mu}\omega t}\, \mathrm{d}t \right] \mathrm{d}\tau$$

we recognize the inner integral as a right-sided transform and applying the shift theorem from [3.16], we can replace it with $\mathcal{F}^R\{g(t)\} e^{\boldsymbol{\mu}\omega\tau}$:

$$r(\omega) = \int_{-\infty}^{\infty} \overline{f(\tau)} \mathcal{F}^R\{g(t)\} e^{\boldsymbol{\mu}\omega\tau} \mathrm{d}\tau$$

Now the problem is one of ordering. We need to place the exponential that we have just inserted adjacent to $\overline{f(\tau)}$. This cannot be done without a decomposition because the exponential does not commute with $\mathcal{F}^R\{g(t)\}$. However, by decomposing $\mathcal{F}^R\{g(t)\}$ into components parallel and perpendicular to $\boldsymbol{\mu}$ using [3.21], it becomes possible to change the order, in one case requiring a conjugate.

$$\begin{aligned} r(\omega) &= \int_{-\infty}^{\infty} \overline{f(\tau)} \left(\mathcal{F}_{\parallel}^R\{g(t)\} + \mathcal{F}_{\perp}^R\{g(t)\} \right) e^{\boldsymbol{\mu}\omega\tau} \mathrm{d}\tau \\ &= \int_{-\infty}^{\infty} \overline{f(\tau)} \mathcal{F}_{\parallel}^R\{g(t)\} e^{\boldsymbol{\mu}\omega\tau} \mathrm{d}\tau + \int_{-\infty}^{\infty} \overline{f(\tau)} \mathcal{F}_{\perp}^R\{g(t)\} e^{\boldsymbol{\mu}\omega\tau} \mathrm{d}\tau \end{aligned} \qquad [4.18]$$

The parallel component commutes with the exponential, whereas the perpendicular component anti-commutes: when the order is changed, the exponential must be conjugated:

$$r(\omega) = \int_{-\infty}^{\infty} \overline{f(\tau)} e^{\boldsymbol{\mu}\omega\tau} \mathrm{d}\tau \, \mathcal{F}_{\parallel}^{R}\{g(t)\} + \int_{-\infty}^{\infty} \overline{f(\tau)} e^{-\boldsymbol{\mu}\omega\tau} \mathrm{d}\tau \, \mathcal{F}_{\perp}^{R}\{g(t)\}$$

The remaining integrals can now be recognized as right-sided transforms, in one case with a negated transform axis, hence:

$$r(\omega) = \mathcal{F}^{R}\left\{\overline{f(t)}\right\} \mathcal{F}_{\parallel}^{R}\{g(t)\} + \mathcal{F}_{-\boldsymbol{\mu}}^{R}\left\{\overline{f(t)}\right\} \mathcal{F}_{\perp}^{R}\{g(t)\} \quad [4.19]$$

and $c(t) = \mathcal{F}^{-R}\{r(\omega)\}$. Note that we have arrived at a result using only right-sided transforms, all in the same direction (forward), but in one case with a negated axis. Only one inverse transform is required as the final step. This means that if there is asymmetry in the scale factors between the forward and inverse transforms, this formula requires no modification, as it would if we mixed forward and inverse transforms.

It remains to explain the meaning of the quaternion values in the correlation result, since it is a quaternion-valued function. It was shown in 2003 [MOX 03] that the quaternion values represent an averaged rotation that in some sense is a best fit rotation to map the quaternion values in one signal into the values in the other signal for each offset between the signals. This value has a more precise meaning when there is a clear correlation "peak", the position of which identifies the offset that best matches one signal to the other.

4.2.3. *Quaternion phase correlation*

Phase correlation works well on images because of the presence of sharp edges (in most images). Therefore, we present here an example of color image phase correlation. Figure 4.12, provided in full color in the color section within this book, shows two images with identical backgrounds.

Each image contains a car, of identical model, but different colors (one is red and the other is yellow). The two cars are placed at different positions in each image (approximately displaced, but also slightly rotated relative to the camera). Figure 4.13, provided in full color in the color section within this book, shows results from phase correlation based on the same operational formula as was given in the previous section, apart from normalization of each value in the cross-power spectrum by dividing by its modulus (exactly as in the complex case and subject to the same provisos about very small values noted immediately after [4.16]):

$$p = \mathcal{F}^{-R}\left\{\frac{r(\omega)}{|r(\omega)|}\right\} \qquad [4.20]$$

where $r(\omega)$ is as given in [4.19]. The maximum amplitude peak in the phase correlation surface is found, and its coordinates used as an offset to shift (cyclically) the second car image. When added to the first car image (and normalized by dividing by 2), the result in Figure 4.13 is obtained. Note that the two cars have been superimposed correctly, although a small rotation is apparent (the two cars do not align precisely). This result is achieved despite the complex background of the image (which admittedly is invariant between the two images). As can be seen from the phase correlation surface, correct detection of the offset between the two cars is only just achieved. There are 20 points with a magnitude greater than 75% of the amplitude of the largest peak, showing that the detection of the offset is close to being incorrect. It was shown in 1998 that complex phase correlation working on chrominance data could also achieve a phase correlation result for this pair of images [THO 98], and the result is in agreement with that obtained above.

Figure 4.12. *Two color images with identical backgrounds but a foreground object (the car) in different positions with different colors. (Images by and reproduced with the permission of Dr. A.L. Thornton [THO 98]) (see color section within this book for a full color version of this figure)*

Tests with grayscale versions of the two images showed that classical phase correlation was unable to establish the offset between the two cars. More than 6,000 points in the phase correlation surface had magnitudes greater than the point at the correct offset of (4,169) found in the quaternion/color case.

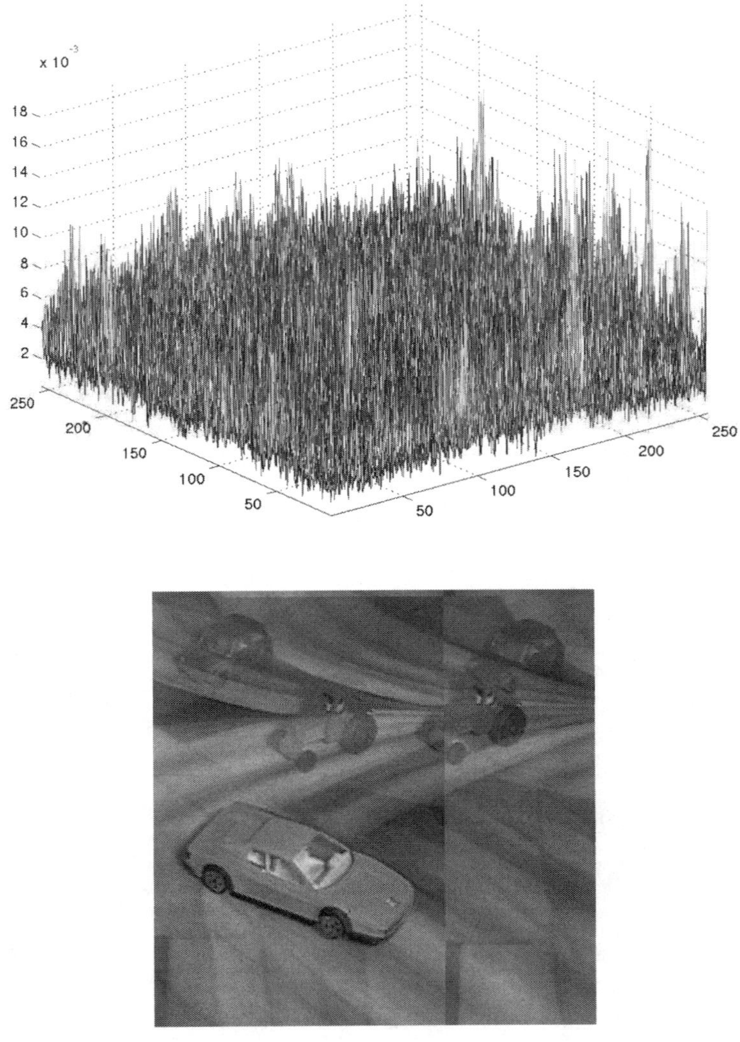

Figure 4.13. *Phase correlation results using the color images in Figure 4.12. Upper: the phase correlation surface. Lower: the two images superimposed after shifting one of them by the coordinate offset found from the maximum peak in the phase correlation surface, showing the two cars correctly superimposed (see color section within this book for a full color version of this figure)*

In a practical application, a better result would be obtained by removing the background from a "reference" image of the car so that the extraneous detail in the reference image would not correlate with details in the "target" image.

Quaternion phase correlation has been shown to work when correlating a color image with a grayscale image [SAN 01], indicating that there could be applications in image registration (e.g. to register an aerial photograph with a map).

4.3. Instantaneous phase and amplitude of complex signals

The notions of instantaneous phase and instantaneous amplitude have been well-known concepts in the signal and image areas for a long time [GAB 46]. A very intuitive parallel, which was made in the early years of time-frequency analysis [VIL 48] by physicists, is to link instantaneous phase to the phase velocity and instantaneous amplitude to the group velocity. Instantaneous phase and amplitude can be considered as the simplest versions of a time-frequency analysis for single frequency signals [FLA 98].

In this section, we show how it is possible to define an instantaneous phase and amplitude for a complex-valued signal using similar techniques as those used for real-valued signals, but by means of quaternion rather than complex Fourier transforms. The work presented here can be seen as an alternate approach to the bivariate signal analysis developed by Lilly *et al.* [LIL 10], where the authors used vector and matrix algebra together with complex Fourier transforms to analyze bivariate (or equivalently complex-valued) signals. Here, the proposed approach is to stick to scalar (univariate) complex-valued signals and use the QFTs to analyze them. The presented results are then straightforward generalizations of existing results for real-valued signals, which become a special case of the theory developed in the following.

In [LIL 10], the authors analyzed complex-valued signals $z(t)$ as bivariate signals that can be represented as trajectories in the plane, as displayed in Figure 4.14. Here, we will rather consider complex signals as trajectories in 3D space, where the third dimension is time t. This is illustrated in Figure 4.15.

This choice for a 3D trajectory representation will become obvious when defining the instantaneous phase and frequency concepts in sections 4.3.4 and 4.3.5 as they will be associated with the 3D behavior of the complex signal $z(t)$.

4.3.1. *Important properties of 1D QFT of a complex signal* $z(t)$

Recall that here we consider complex signals $z(t) = z_r(t) + \boldsymbol{i}z_i(t)$, with real part $z_r(t)$ and imaginary part $z_i(t)$, which belong to $L^1(\mathbb{R};\mathbb{C}) \cap L^2(\mathbb{R};\mathbb{C})$. Such complex signals have a right-sided QFT with axis $\boldsymbol{j}$, denoted as $Z_{\boldsymbol{j}}^R(\omega)$, given by:

$$Z_{\boldsymbol{j}}^R(\omega) = \mathcal{F}_{\boldsymbol{j}}^R\{z\}(\omega) = \int_{-\infty}^{+\infty} z(t)e^{-\boldsymbol{j}\omega t}\,\mathrm{d}t \qquad [4.21]$$

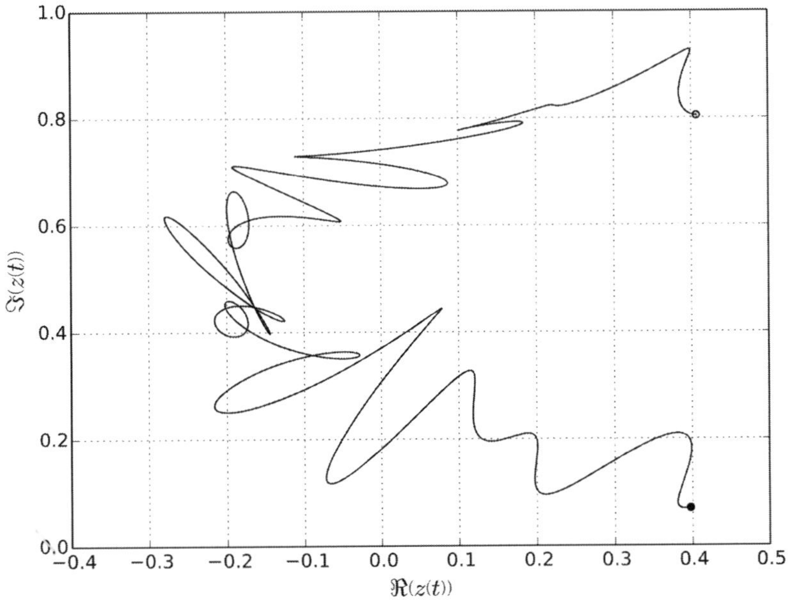

Figure 4.14. *Representations of a complex signal $z(t)$ as a trajectory in a 2D plane with constant time step from its initial point (black dot) at $t=0$ to its final point (black circle) at $t=1$*

Note that we consider a right-sided QFT here, and that similar results could be obtained by choosing a left-sided QFT. Also, in this section, we take the axis of the transform to be $\boldsymbol{\mu} = \boldsymbol{j}$, a special case of the general definition of the right-sided QFT given in [3.5].

Now, given the quaternion-valued Fourier transform with respect to axis $\boldsymbol{j}$, denoted $Z_{\boldsymbol{j}}(\omega) \in L^1(\mathbb{R};\mathbb{H}) \cap L^2(\mathbb{R};\mathbb{H})$, then its inverse transform is:

$$z(t) = \int_{-\infty}^{+\infty} Z_{\boldsymbol{j}}(\omega) e^{\boldsymbol{j}\omega t}\, \mathrm{d}\omega \qquad [4.22]$$

where $z(t)$ is in general complex-valued. Note that in some cases to be discussed later on, the signal so constructed $z(t)$ may be quaternion-valued, depending on the values taken by $Z_{\boldsymbol{j}}(\omega)$. Now, we present some properties of the QFT that will be useful in the definition of instantaneous attributes. The key property that allows a quaternion extension to what is known for a real signal and its complex-valued analytic signal comes from the symmetry properties of the QFT.

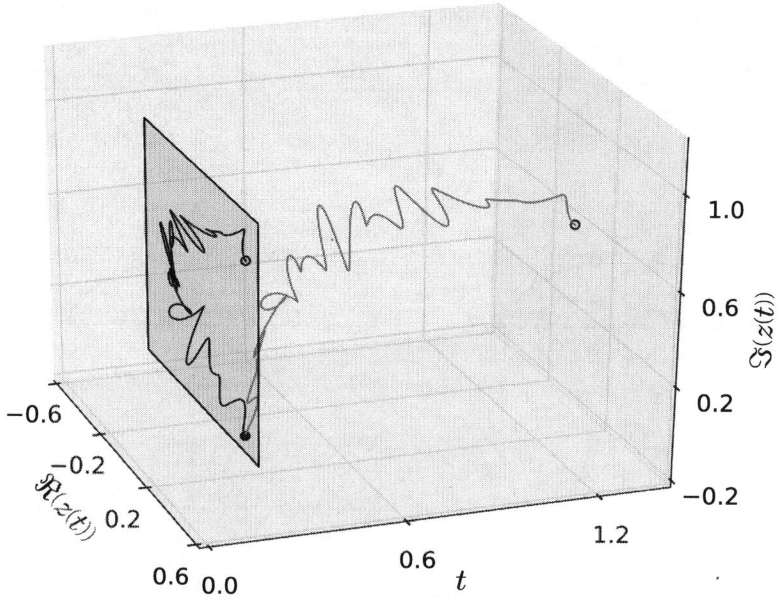

Figure 4.15. *2D and 3D representations of a complex signal z(t). The 2D trajectory in the plane (black signal) is the same as in Figure 4.14. The 3D trajectory (gray signal) is obtained by unfolding z(t) along a third dimension, which corresponds to t. In the 3D representation, each point has coordinate* $[\Re(z(t)), \Im(z(t)), t]$. *The initial point (black dot) is displayed at time t = 0.2 and the final point (black circle) is displayed at time t = 1.2 (the 0.2 shift is used for display convenience)*

4.3.1.1. *Symmetry*

Given a real-valued signal $x(t)$, it is well known that its (complex) Fourier transform fulfills Hermitian symmetry, such that $X(-\omega) = X^*(\omega)$, where * denotes the complex conjugate. Now, it is also well known that the (complex) Fourier transform of a complex-valued signal $z(t)$ has no symmetry. This lack of symmetry suggests that an alternate Fourier transform may be used that possesses some symmetry properties. This is where the QFT can be the alternative to a classical Fourier transform.

The QFT with axis $\boldsymbol{j}$ of a complex signal $z(t) \in \mathbb{C}_{\boldsymbol{i}}$, denoted as $Z_{\boldsymbol{j}}(\omega)$, is $\boldsymbol{i}$-*Hermitian symmetric*, i.e.:

$$Z_{\boldsymbol{j}}(\omega) = \overline{Z_{\boldsymbol{j}}(-\omega)}^{\boldsymbol{i}} \qquad [4.23]$$

Figure 4.16 illustrates this symmetry property.

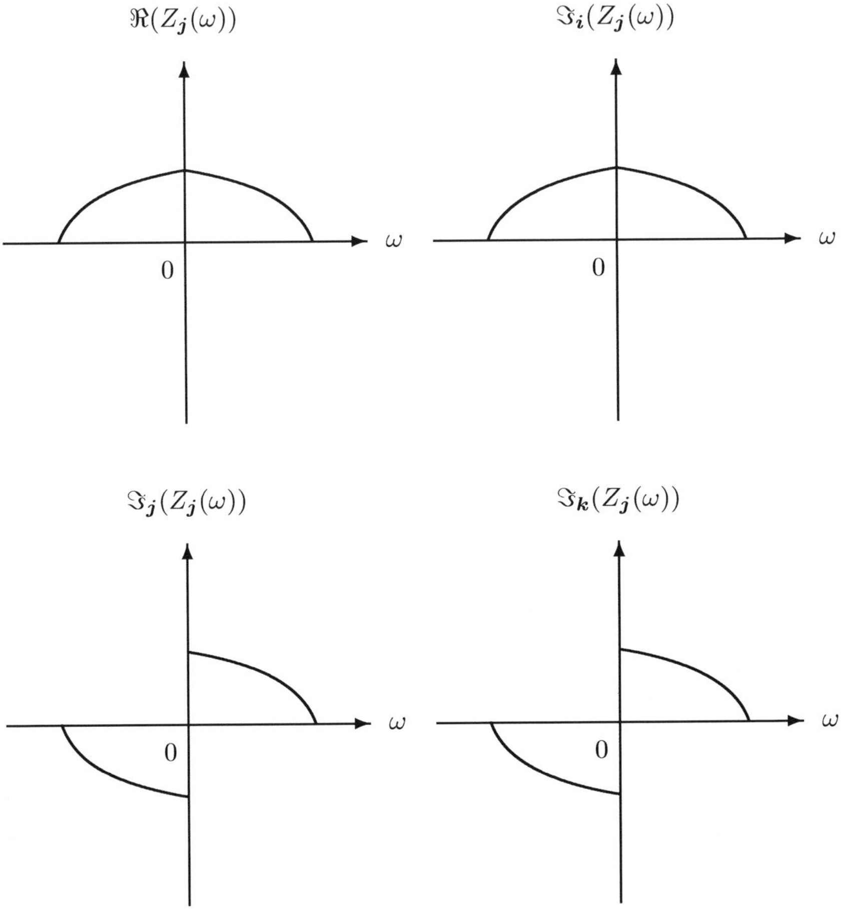

Figure 4.16. *Symmetry of the four components of the QFT $Z_j(\omega)$ of a complex signal $z(t)$. The QFT satisfies the i-Hermitian symmetry: $Z_j(\omega) = \overline{Z_j(-\omega)}^{i}$, which means that $\Re(Z_j(\omega))$ and $\Im_i(Z_j(\omega))$ are even while $\Im_j(Z_j(\omega))$ and $\Im_k(Z_j(\omega))$ are odd*

An additional symmetry property is also of interest when considering complex signals $z(t)$. The QFT of the conjugate of z is given by:

$$\mathcal{F}_j^R\{z^\star\}(\omega) = \overline{\mathcal{F}_j^R\{z\}(\omega)}^{j} \qquad [4.24]$$

Note that good symmetries do not exist for the (classical) complex Fourier transform of a complex signal and that the QFT allows us to define a Fourier transform with symmetries for complex signals.

4.3.1.2. *Orthocomplex modulation*

A key property of the 1D QFT is its behavior with respect to modulation. It is well known that the classical (complex) Fourier transform of a modulated signal consists of a frequency-shifted version of the unmodulated version of the signal. That is, for a given real or complex signal $z(t)$ with FT $Z(\omega)$, the FT of $z(t)e^{i\omega_0 t}$ is $Z(\omega - \omega_0)$ for a given constant frequency ω_0. For the QFT, things are different when one considers a complex-valued signal $z(t)$. This can be seen by inserting $w(t) = z(t)e^{i\omega_0 t}$ in place of $z(t)$ in [4.21] which gives:

$$\mathcal{F}_{\boldsymbol{j}}\{w\}(\omega) = W_{\boldsymbol{j}}(\omega) = \int_{-\infty}^{+\infty} w(t)e^{-\boldsymbol{j}\omega t}\,\mathrm{d}t = \int_{-\infty}^{+\infty} z(t)e^{-\boldsymbol{i}\omega_0 t}e^{-\boldsymbol{j}\omega t}\,\mathrm{d}t \qquad [4.25]$$

Now, recall from [1.38] that:

$$e^{\boldsymbol{i}\alpha}e^{-\boldsymbol{j}\beta} \neq e^{(\boldsymbol{i}\alpha - \boldsymbol{j}\beta)} \qquad [4.26]$$

for any $\alpha, \beta \in \mathbb{R}$, then the property of modulation does not extend trivially to the 1D QFT case. In fact, the QFT of $w(t)$ reads:

$$W_{\boldsymbol{j}}(\omega) = \frac{1}{2}\left(Z_{\boldsymbol{j}}(\omega - \omega_0) + Z_{\boldsymbol{j}}(\omega + \omega_0) - \boldsymbol{i}Z_{\boldsymbol{j}}(\omega - \omega_0)\boldsymbol{j} + \boldsymbol{i}Z_{\boldsymbol{j}}(\omega + \omega_0)\boldsymbol{j}\right) \quad [4.27]$$

which cannot be considered as a modulation in the sense that several copies of the QFT of $z(t)$ are shifted and modified.

In order to recover the same frequency shift behavior as in the well-known complex case, it is necessary to use another axis for the exponential that multiplies the signal. In fact, the well-known phase shift appears when the axis of the exponential is the same as the transform axis. So now, consider the signal $y(t) = z(t)e^{-\boldsymbol{j}\omega_0}$, then its QFT (with axis $\boldsymbol{j}$), denoted as $Y_{\boldsymbol{j}}(\omega) = \mathcal{F}_{\boldsymbol{j}}\{y\}(\omega)$, is given by:

$$Y_{\boldsymbol{j}}(\omega) = \int_{-\infty}^{+\infty} z(t)e^{-\boldsymbol{j}(\omega - \omega_0)t}\,dt = Z_{\boldsymbol{j}}(\omega - \omega_0) \qquad [4.28]$$

Note also that a special case of orthocomplex modulation is that the QFT with axis $\boldsymbol{j}$ of $\exp(\boldsymbol{j}\omega_0)$ is simply $\delta(\omega - \omega_0)$. This orthocomplex modulation property extends to QFTs with any axis $\boldsymbol{\mu}$, provided that the exponential axis is also $\boldsymbol{\mu}$.

Note that the orthocomplex modulated signal $y(t) = z(t)e^{-\boldsymbol{j}\omega_0}$ obtained by orthocomplex modulation of the complex signal $z(t)$, is quaternion-valued, i.e. $y(t) \in \mathbb{H}$. Thus, when taking a base-band complex signal and orthocomplex modulating it, we get a quaternion-valued signal. As will also be shown in section 4.3.3, a signal with a right-sided QFT is quaternion-valued in the time domain. This will be exploited to propose the quaternion extension of a complex signal in section 4.3.3.

4.3.1.3. *Convolution*

Here, we are interested in the convolution of two complex-valued signals. The very interesting property that exists for such a convolution reads as follows. Consider two complex signals $z(t)$ and $w(t)$ with QFTs $Z_{\boldsymbol{j}}(\omega)$ and $W_{\boldsymbol{j}}(\omega)$, respectively, then the following stands:

$$\mathcal{F}_{\boldsymbol{j}}\left\{z * w\right\}(\omega) = Z_{\boldsymbol{j}}(\omega) \odot W_{\boldsymbol{j}}(\omega) \qquad [4.29]$$

where $\odot$ denotes the bicomplex product of [1.65].

Note that the *bicomplex product* (defined in 1.4.2) is commutative so that, as one would hope, the order in the convolution product does not affect the result, i.e.:

$$\mathcal{F}_{\boldsymbol{j}}\left\{z * w\right\}(\omega) = \mathcal{F}_{\boldsymbol{j}}\left\{w * z\right\}(\omega) \qquad [4.30]$$

for any two complex-valued signals $z(t)$ and $w(t)$. This property is rather important as it allows us to deal with convolution products of complex signals and their QFTs just as we are used to doing with real signals and their complex FTs. This simply means that the presented results are generalizations of existing facts and that the QFT offers a thorough generalization of known results to deal with complex-valued signals.

4.3.2. *Hilbert transform and right-sided quaternion spectrum*

As we are interested in defining instantaneous amplitude and phase for complex signals, a simple way to do it is to mimic the Gabor [GAB 46] and Ville [VIL 48] approach that adds a signal in quadrature to the original signal. In their work, Gabor and Ville considered real-valued signals and added a quadrature signal in the imaginary part of a complex-valued signal associated with the original. This complex signal is named the "analytic signal" associated with a real-valued signal. The quadrature signal was obtained by the Hilbert transform of the original signal.

Here, our original signal is complex-valued. So, in order to proceed in the manner of Gabor and Ville, we now need to add a complex signal in quadrature to the first one and to make this addition so that the final signal is 4D. This is where quaternion-valued signals enter the scene. Before building this quaternion signal, we now take a look at how to build the quadrature signal. For this, we would like to simply build the Hilbert transform of a given complex signal $z(t)$. It is constructed as follows.

First, we are interested in the QFT of the real-valued signal $x(t) = 1/\pi t$. It is trivial to notice that, due to the behavior of the QFT with respect to real-valued signals (see Chapter 3), we get:

$$\mathcal{F}_{\boldsymbol{j}}\{x\}(\omega) = -\boldsymbol{j}\,\mathrm{sgn}(\omega) \tag{4.31}$$

This allows us to define, in a straightforward way, the Hilbert transform of a complex signal $z(t) \in \mathbb{C}$, denoted $z_{\mathcal{H}}(t)$, as follows:

$$z_{\mathcal{H}}(t) = p.v. \left(z * \frac{1}{\pi t}\right)(t) \tag{4.32}$$

Now, as $1/\pi t$ is a real-valued signal, if we look at the QFT of $z_{\mathcal{H}}(t)$, it is the QFT of the convolution product between a real and a complex signal. Thus, the previously introduced property [4.29] for the transformation of a convolutive product applies in a degenerate form where one of the two signals has no imaginary part. In this case, as can be checked directly from [4.29], the *bicomplex product* boils down to a classical product and reads:

$$Z_{\boldsymbol{j}\mathcal{H}}(\omega) = -\boldsymbol{j}\,\mathrm{sgn}(\omega) \odot Z_{\boldsymbol{j}}(\omega) = -\boldsymbol{j}\,\mathrm{sgn}(\omega) Z_{\boldsymbol{j}}(\omega) \tag{4.33}$$

A proof can be found in [LE 14]. As we have defined the *quadrature* signal $z_{\mathcal{H}}(t)$ of a given complex signal $z(t)$, we need to check which type of signal corresponds to a spectrum with only positive frequencies. Consider a spectrum $Z_{\boldsymbol{j}}(\omega)$ obtained from the QFT of a complex signal $z(t) = z_r(t) + \boldsymbol{i} z_i(t)$. Then, just as shown in Figure 4.16, the spectrum $Z_{\boldsymbol{j}}(\omega)$ has positive and negative frequencies. In order to build from $Z_{\boldsymbol{j}}(\omega)$ a spectrum with only positive frequencies, we simply need to consider $(1 + \mathrm{sgn}(\omega))\, Z_{\boldsymbol{j}}(\omega) = W_{\boldsymbol{j}}(\omega)$. It is obvious that $W_{\boldsymbol{j}}(\omega) = 0, \quad \forall \omega < 0$, and by linearity of the inverse QFT, we get:

$$\begin{aligned}
\mathcal{F}_{\boldsymbol{j}}^{-1}\{W_{\boldsymbol{j}}\}(t) &= \mathcal{F}_{\boldsymbol{j}}^{-1}\{(1 + \mathrm{sgn}(.))\, Z_{\boldsymbol{j}}\}(t) \\
&= \mathcal{F}_{\boldsymbol{j}}^{-1}\{Z_{\boldsymbol{j}}\}(t) + \mathcal{F}_{\boldsymbol{j}}^{-1}\{\mathrm{sgn}(.) Z_{\boldsymbol{j}}\}(t) \\
&= \mathcal{F}_{\boldsymbol{j}}^{-1}\{Z_{\boldsymbol{j}}\}(t) + \boldsymbol{j}\mathcal{F}_{\boldsymbol{j}}^{-1}\{-\boldsymbol{j}\,\mathrm{sgn}(.) Z_{\boldsymbol{j}}\}(t)
\end{aligned} \tag{4.34}$$

where the last equality is valid due to the fact that we are considering a *right*-sided QFT. From the last equality in [4.34] and equation [4.33], we can see that the signal $w(t) = \mathcal{F}_{\boldsymbol{j}}^{-1}\{W_{\boldsymbol{j}}\}(t)$ is simply the following combination:

$$w(t) = z(t) + \boldsymbol{j} z_{\mathcal{H}}(t) \quad [4.35]$$

which is the sum of the original signal $z(t)$ and its *quadrature* version that is put in two extra dimensions spanned by the basis vectors $\boldsymbol{j}$ and $\boldsymbol{k}$. As a result, we can see that a one-sided spectrum QFT signal $w(t)$ is indeed a quaternion-valued signal, that is $w(t) \in \mathbb{H}$. This is the direct counterpart of what is known for the classical FT, where a one-sided FT corresponds to a complex-valued signal.

As an example, we mention here what happens when the signal $z(t)$ is band-limited and then orthocomplex modulated. This type of signal was introduced previously and referred to as an *orthocomplex modulated signal.* So, consider a signal $z(t)$ with a band-limited QFT, i.e. $Z_{\boldsymbol{j}}(\omega) = 0$ for $|\omega| > \omega_m$ for a given ω_m. Then, for a frequency $\omega_0 > \omega_m$, the orthocomplex modulated signal $z(t)e^{\boldsymbol{j}\omega_0 t}$ has a right-sided QFT, and, as a result, should be quaternion-valued. This can be directly checked from its literal expression as:

$$\begin{aligned} z(t)e^{\boldsymbol{j}\omega_0 t} &= (z_r(t) + \boldsymbol{i}z_i(t))\left(\cos(\omega_0 t) + \boldsymbol{j}\sin(\omega_0 t)\right) \\ &= z_r(t)\cos(\omega_0 t) + \boldsymbol{i}z_i(t)\cos(\omega_0 t) \\ &\quad + \boldsymbol{j}z_r(t)\sin(\omega_0 t) + \boldsymbol{k}z_i(t)sin(\omega_0 t) \end{aligned} \quad [4.36]$$

Illustration of this orthocomplex modulated signal $z(t)e^{\boldsymbol{j}\omega_0}$ is given in Figure 4.17, where the QFT spectrum $Z_{\boldsymbol{j}}(\omega - \omega_0)$ is drawn.

We now move to the problem of associating a quaternion-valued signal with any complex signal, making use of the behavior of the QFT just presented.

Even	Odd
$\Re(Z_{\boldsymbol{j}}(\omega))$	$\Im_{\boldsymbol{j}}(Z_{\boldsymbol{j}}(\omega))$
$\Im_{\boldsymbol{i}}(Z_{\boldsymbol{j}}(\omega))$	$\Im_{\boldsymbol{k}}(Z_{\boldsymbol{j}}(\omega))$

Table 4.1. *Symmetry properties of the components of the QFT $Z_{\boldsymbol{j}}(\omega)$ of a complex signal $z(t)$*

4.3.3. *The quaternion signal associated with a complex signal*

Due to the results presented in the previous section 4.3.2, we can now turn to the definition of a quaternion-valued signal associated with any complex signal $z(t)$. We have just seen that a quaternion-valued signal corresponds to a one-sided spectrum. It seems now rather clear how to associate a given complex signal $z(t)$ with its quaternion companion. However, one could imagine several ways to make such an association. The one presented here is similar to the association between a real signal and its associated analytic signal: they share the same spectral content. This is explained in detail in [PIC 97], from which we just recall the main idea and follow the lines.

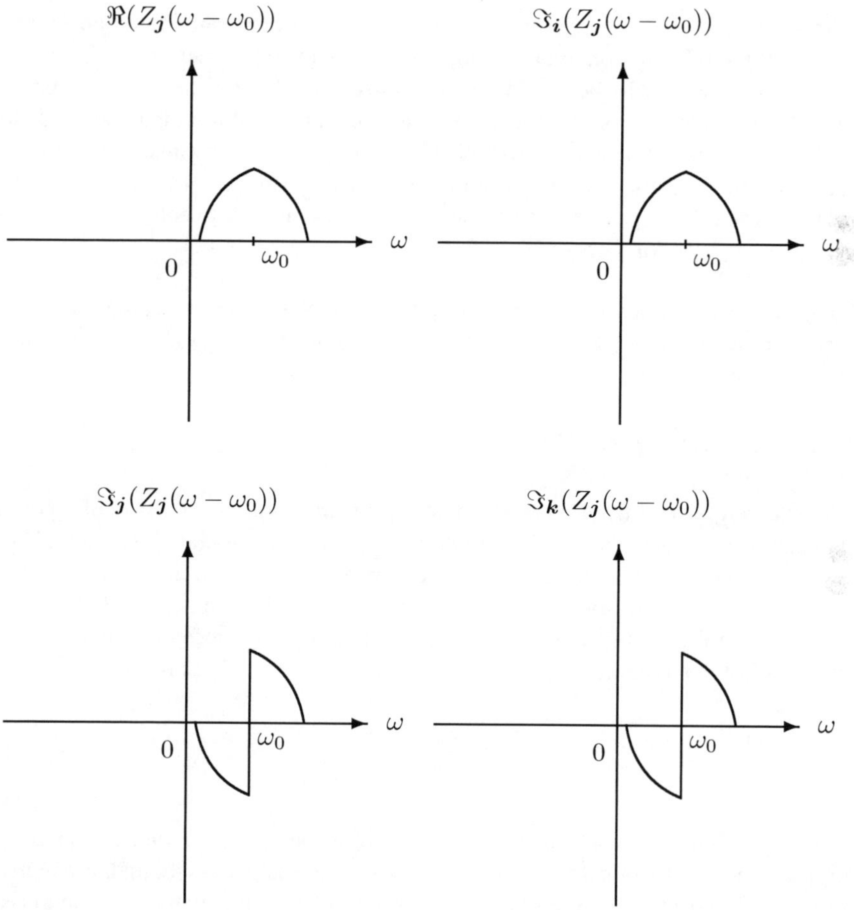

Figure 4.17. *Four components of the quaternion-valued Fourier transform $Z_{\boldsymbol{j}}(\omega - \omega_0)$ of the orthocomplex modulated signal $z(t)e^{\boldsymbol{j}\omega_0 t}$ and where the complex signal $z(t)$ has a QFT verifying $|Z_{\boldsymbol{j}}(\omega)| = 0$ for $|\omega| > \omega_0$. The spectrum of $z(t)$ is displayed with the standard symmetries holding for a QFT (with axis $\boldsymbol{j}$) of complex signals*

Given a signal $z(t)$, there is an infinite number of choices of corresponding quaternion-valued signals $q_z(t)$. In order to uniquely determine this signal, the most natural choice is to ask that the frequency content of the signal should be the same as that of $z(t)$. Now, remembering the symmetry properties of the QFT depicted in Figure 4.16, it is obvious that only half of the spectrum (positive or negative frequencies) of $z(t)$ is necessary to reconstruct the entire spectrum due to the symmetry properties of the QFT (with axis $\boldsymbol{j}$) of $z(t)$.

As a consequence, we can simply choose to keep the right-sided spectrum that contains all the information about $z(t)$. As we just showed in section 4.3.2, the signal constructed as such will be a quaternion-valued signal (obtained, for example, by inverse QFT of the right-sided spectrum), and this quaternion-valued signal, denoted as $q_z(t)$ previously, will share exactly the same frequency properties as the "original" signal $z(t)$. In fact, this way of building a quaternion-valued signal from a complex signal allows us to make a one-to-one correspondence between the two signals (through the content of their QFTs).

To summarize, if we are given a complex signal $z(t)$, then we can associate with it a quaternion signal $q_z(t)$ that we will call the *quaternion extension* of $z(t)$ and that is simply constructed as:

$$q_z(t) = z(t) + \boldsymbol{j}\mathcal{H}_{\boldsymbol{j}}\{z\}(t) \qquad [4.37]$$

As can be deduced from what we exposed, the *quaternion extension* of a complex signal can be seen as the equivalent of the analytic signal associated with a real signal as first introduced by Gabor [GAB 46]. The property of analyticity is also shared by $q_z(t)$ in a certain way. In fact, the quaternion extension can be called *pair-wise analytic* in the following sense. Consider $q_z(t) \in \mathbb{H}$, a quaternion extension signal corresponding to a complex signal $z(t)$, with the following $\mathbb{C}_{\boldsymbol{j}}$-pair form:

$$\begin{aligned} q_z(t) &= \left(\Re\left(q_z(t)\right) + \boldsymbol{j}\Im_{\boldsymbol{j}}\left(q_z(t)\right), \Im_{\boldsymbol{i}}\left(q_z(t)\right) + \boldsymbol{j}\Im_{\boldsymbol{k}}\left(q_z(t)\right)\right) \\ &= \left(q_1(t), q_2(t)\right) \end{aligned} \qquad [4.38]$$

Then, the quaternion extension $q_z(t)$ is said to be *pair-wise analytic* as the two complex-valued signals $q_1(t)$ and $q_2(t)$ are analytic signals associated, respectively, with the real and imaginary parts of $z(t)$. Analyticity is understood in the classical sense here, with the slight difference that the complex signals considered are $\mathbb{C}_{\boldsymbol{j}}$-valued. This is a minor issue as $\mathbb{C}_j$ is isomorphic to the classical $\mathbb{C}$ field. The demonstration of the *pair-wise analyticity* of $q(t)$ is trivial. It can be guessed easily by understanding how the QFT acts on a complex signal $z(t)$. The real and imaginary parts of the signal are transformed separately and the $\mathbb{H}$-valued spectrum obtained by a QFT can be seen as a pair of Fourier transforms applied to the real and imaginary parts of $z(t)$. The $\Re$ and $\Im_{\boldsymbol{j}}$ parts of the QFT contain information from the real part of $z(t)$ and the $\Im_{\boldsymbol{i}}$ and $\Im_{\boldsymbol{k}}$ parts contain information from the imaginary part of $z(t)$. Then, by linearity of the Hilbert transformation, we can see that the process of constructing the *quaternion extension* consists of processing the information contained in the real and imaginary parts of $z(t)$ in parallel. In other words, the *quaternion extension* of a complex signal $z(t)$ consists of associating with it an *analytic pair* of signals with components that are the analytic signals associated with the real and imaginary parts of $z(t)$. These analytic signals are then the components of the $\mathbb{C}_{\boldsymbol{j}}$-pair form of the *quaternion extension* $q_z(t)$.

4.3.4. *Instantaneous amplitude and phase*

We have just constructed a quaternion signal $q_z(t)$ that is defined in a unique way from a complex signal $z(t)$, following the tracks of the construction of the analytic signal associated with a real signal. The natural question to ask now is:

> Can we obtain an instantaneous amplitude and phase of a complex signal $z(t)$ from its quaternion extension $q_z(t)$, and if so, what do the amplitude and phase represent?

In the analytic signal case, provided that the spectral content of the signal is "simple",[7] the analytic signal is understood as the simplest time-frequency representation. Now it is no longer the best representation to use if the spectral content of the signal is richer. In such cases, we should refer to more sophisticated time-frequency analyses (e.g. see the Wigner-Ville representation in section 4.3.7). However, the links between the analytic signal and time-frequency representations (such as the Wigner-Ville distribution) are well known [VIL 48]. They are investigated for the quaternion case in sections 4.3.8 and 4.3.9.

From what was explained in section 4.3.3, one might think that the most natural way to handle the *quaternion extension* $q_z(t)$ is to consider it as a pair of complex signals. It is tempting to define a pair of instantaneous amplitudes as well as a pair of instantaneous phases. This was done by Lilly *et al.* [LIL 10] and extended to signals with more than two components in [LIL 11]. However, here we would like to consider the signal $q_z(t)$ as a single entity rather than analyzing it component-wise. This is because we would like to take advantage of the geometric nature of quaternions. To do so, it is possible to use the richness of quaternion representations/forms to express the quaternion extension signal in terms of a modulus and a phase. Clearly, there are two polar forms that could be used: the polar and the CD polar form. We will use the latter here because it includes a complex-valued envelope which is of particular interest in the analysis of a complex signal.

7 Here, "simple" means that the spectral content is made up of one bandpass contribution.

Given a complex signal $z(t)$ and its *quaternion extension* $q_z(t)$, there is a *canonical pair* $[\rho_z(t), \phi_z(t)]$, with $\rho_z(t) \in \mathbb{C}$ and $\phi_z(t) \in \mathbb{C}$ uniquely associated with $z(t)$ by the polar CD form of $q_z(t)$. Recall that the quaternion extension $q_z(t)$ possesses the same spectral information as $z(t)$ and that it is made up of the signal $z(t)$ and its orthogonal Hilbert transform with respect to the $\boldsymbol{j}$ axis (see [4.37]). Then, using the polar CD form introduced in 1.4.1.5, the quaternion extension $q_z(t)$ is:

$$q_z(t) = \rho_z(t)e^{\phi_z(t)\boldsymbol{j}} \qquad [4.39]$$

The principle of canonical pairs (modulus and phase) associated with a given signal was originally used by Picinbono [PIC 97] to demonstrate the one-to-one relation between a real signal and its instantaneous amplitude and phase defined through the analytic signal [VIL 48].

There is a one-to-one correspondence between a complex signal $z(t)$ and the canonical pair associated with $z(t)$ through its quaternion extension $q_w(t)$ (see details in [LE 14]).

Now, looking at the polar CD form of the quaternion extension $q_z(t)$ in [4.39], it is noticeable that it looks very much like an orthocomplex modulation. Recall that the quaternion extension has, by construction, no negative frequencies. As a consequence, it can be shown (see [LE 14] for details) that $\phi_z(t)$ is not complex but real-valued[8]. Thus, the polar CD form of $q_z(t)$ should be understood as made up of a "low frequency" part $\rho_z(t)$ (the instantaneous amplitude/modulus) subject to an orthocomplex modulation with the exponential $e^{\phi_z(t)\boldsymbol{j}}$ where $\phi_z(t)$ is the instantaneous phase of $z(t)$. Once again, note that this is a very thorough extension of the concepts well known for real signals when considering their complex extension (analytic signal). The presented results incorporate the analytic case as a special case, which happens when $z(t)$ is real-valued. Some illustrative examples of the concept of instantaneous amplitude and phase for different types of complex signals $z(t)$ are presented in section 4.3.6.

4.3.5. *The instantaneous frequency of a complex signal*

Of great interest in non-stationary signal processing is the concept of instantaneous frequency, which provides information on the time-varying spectral content of a signal. However, it must be remembered that this can only lead to a proper interpretation when the spectral content of the signal is not too complicated.

8 Note that for quaternion-valued signals with a one-sided QFT, the argument/phase in their polar CD form is always real. However, there are some quaternion-valued signals having a two-sided QFT spectrum, but with no specific symmetry relations. For such signals, the argument/phase of their polar CD form can be complex-valued.

When such cases arise, the notion of *instantaneous frequency* of a complex signal $z(t)$, denoted by $\Omega_z(t)$, is directly linked to the instantaneous phase defined previously, as:

$$\Omega_z(t) = \frac{\mathrm{d}\phi_z(t)}{\mathrm{d}t} \tag{4.40}$$

Once again, remember that the considered phase is real-valued, leading to a real-valued instantaneous frequency. This definition is again a generalization of what is known for real-valued signals [VIL 48].

Now, recalling that the signal we are looking at, i.e. $z(t)$, is complex-valued and bearing in mind the 2D planar or 3D trajectory that characterizes $z(t)$, it is obvious that we need more than just an instantaneous frequency to analyze such a signal.

Due to the 2D nature of its samples, it is possible to think about $z(t)$ as a trajectory evolving in the 2D plane as mentioned earlier. However, here we prefer to consider $z(t)$ as a 3D curve, as illustrated in Figure 4.15. As we have obtained an instantaneous frequency $\Omega(t)$ for $z(t)$ and as it is scalar (real) valued, there should be more information necessary to describe the instantaneous behavior of $z(t)$. In fact, in addition to the value of the instantaneous frequency $\Omega_z(t)$, we need to know in which plane in 3D space is the signal oscillating at each time t. This notion can be understood as the *instantaneous osculating plane* of $z(t)$.

This osculating plane can be obtained from the instantaneous amplitude introduced in section 4.3.4, as explained in [LE 14]. This can be inferred from the classical case of the analytic signal associated with a real signal $x(t)$. The instantaneous amplitude, also known as the *envelope* of the signal, is obtained from the modulus of the analytic signal. Now, as shown in Figure 4.18, the envelope $\rho_x(t)$ together with its first-order derivative $\rho'_x(t)$ define the plane in which the signal $x(t)$ oscillates. In the case where $x(t)$ is real-valued, it is trivial as the plane does not change with time (the plane is defined by the time axis and the signal magnitude axis).

Now, in the case where the signal $z(t)$ is complex-valued, and remembering that we consider that $z(t)$ is a 3D trajectory (as displayed in Figure 4.15), the envelope is complex-valued, meaning that it can also be interpreted as a 3D trajectory, just like $z(t)$. As a consequence, the derivative of the envelope, $\rho'_z(t)$ is the set of vectors tangent to $\rho_z(t)$, just like the velocity vector associated with a trajectory (see [DO 76] for example). Now, the osculating plane is simply defined by $\rho_z(t) \times \rho'_z(t)$ where $\times$ is the cross product. In the Frenet-Serret formulation [DO 76], the vector defined by this cross product is the *binormal* vector. It is orthogonal to the osculating plane. This osculating plane concept for complex signals will be illustrated in section 4.3.6.

Now, we have shown that in the case of a complex signal $z(t)$, it is possible to define the concept of an instantaneous frequency $\Omega_z(t)$. However, additional information is necessary to describe the dynamical behavior of $z(t)$, and this information is provided by the complex envelope $\rho_z(t)$. First, just as in the real case, it provides a complex envelope to the signal $z(t)$, i.e. an instantaneous amplitude, and in addition to this, it provides the orientation in 3D space of the plane in which $z(t)$ oscillates. The information about this plane and the instantaneous frequency provide complete information about the local behavior of $z(t)$ in 3D. The introduced concepts are illustrated using two simple examples in the following section.

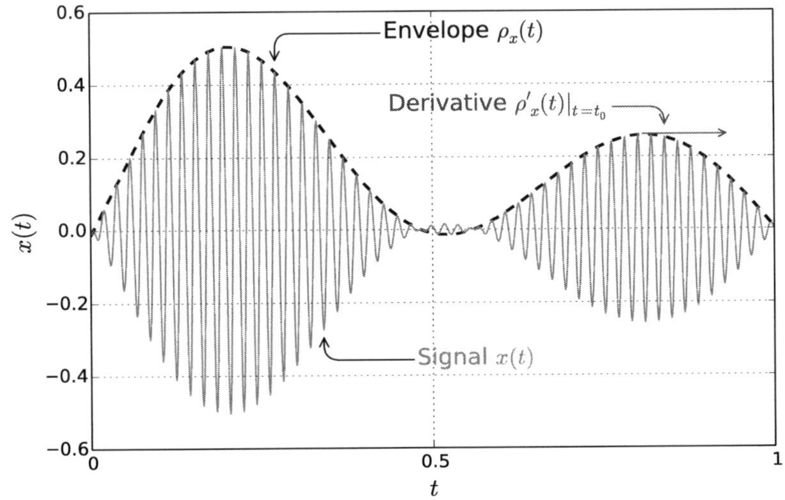

Figure 4.18. *In the classical (real) case when x(t)* $\in \mathbb{R}$ *, the instantaneous amplitude* $\rho_x(t)$ *(dashed black line) of a modulated real signal x(t) (gray) is given by the envelope of its analytic signal. Assuming that* $\rho_x(t)$ *is at least* C^1*, its derivative* $\rho'_x(t)$ *(represented at* $t = t_0$*) together with* $\rho_x(t)$ *defines the osculating plane of the signal x(t)*

4.3.6. *Examples*

In order to show the information carried by the instantaneous phase and amplitude of the quaternion extension associated with a complex signal, we present two toy examples that should highlight the concepts of instantaneous phase and envelope that were previously introduced. The presented examples are orthocomplex modulated complex signals with different kinds of behavior for their instantaneous frequency.

4.3.6.1. *Example 1: orthocomplex modulated signal with constant frequency*

In this example, we consider a complex signal $z(t)$, which is displayed in blue in Figure 4.19 (provided in full color in the color section within this book) together with its real and imaginary parts (projections). This signal $z(t)$ with band-limited QFT spectrum is also displayed (real and imaginary parts) in Figure 4.20. Note that it is not analytic in the classical sense [VIL 48], meaning that its real and imaginary parts are not in quadrature at any time instant. In Figure 4.19, a 3D representation of the signal $z(t)$ is given together with its projections in the $\Re$ and $\Im$ planes. Figure 4.20 presents the projections of the signal, i.e. its real and imaginary parts $\Re(z(t))$ and $\Im((z(t))$.

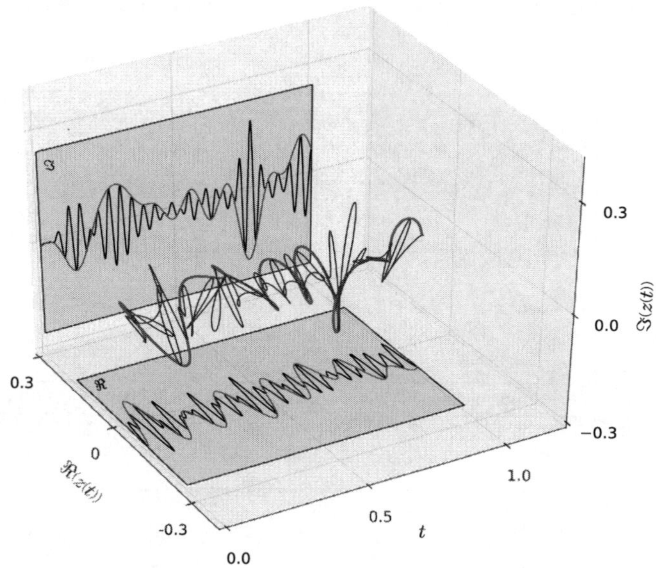

Figure 4.19. *Orthocomplex modulated complex signal with constant frequency $z(t)$. 3D representation of the complex signal $z(t)$ (blue solid line) and its complex instantaneous amplitude $\rho_z(t)$ (red solid line) obtained from the polar CD form of its quaternion extension $q_z(t)$. The projections in $\Re$ and $\Im$ planes (black solid lines) correspond to the signals presented in Figure 4.20 (see color section within this book for a full color version of this figure)*

From the original complex signal $z(t)$, it is possible to estimate its complex envelope $\rho_z(t)$ as explained in section 4.3.4. This is done by computing the QFT of $z(t)$, and then by canceling out the negative frequencies and computing the inverse QFT. The result is a quaternion-valued signal $q_z(t)$. The complex envelope $\rho_z(t)$ of this equivalent quaternion signal $q_z(t)$, obtained from its polar CD form section 1.4.1.5) is displayed in Figures 4.19 and 4.20. One can see from these figures (3D and projections) that $\rho_z(t)$ is a complex-valued envelope. This is thus a complex-valued signal that corresponds to a base-band version of $z(t)$. Now, as explained previously, as $q_z(t)$ is quaternion-valued, its polar CD form is:

$$q_z(t) = \rho_z(t)e^{\Phi_z(t)\boldsymbol{j}} \qquad [4.41]$$

from which we already mentioned the envelope $\rho_z(t)$. In this example, the signal $z(t)$ was constructed such that it contains only one frequency. This can be seen from Figures 4.19 and 4.20 as the oscillations of $z(t)$ are not varying with time, showing a single frequency content behavior. This means that its instantaneous frequency is constant. This could be seen as well by plotting the estimated $\Phi_z(t)$. As it is a rather trivial case, the display is omitted here and an exposition of instantaneous frequency is postponed to example 2.

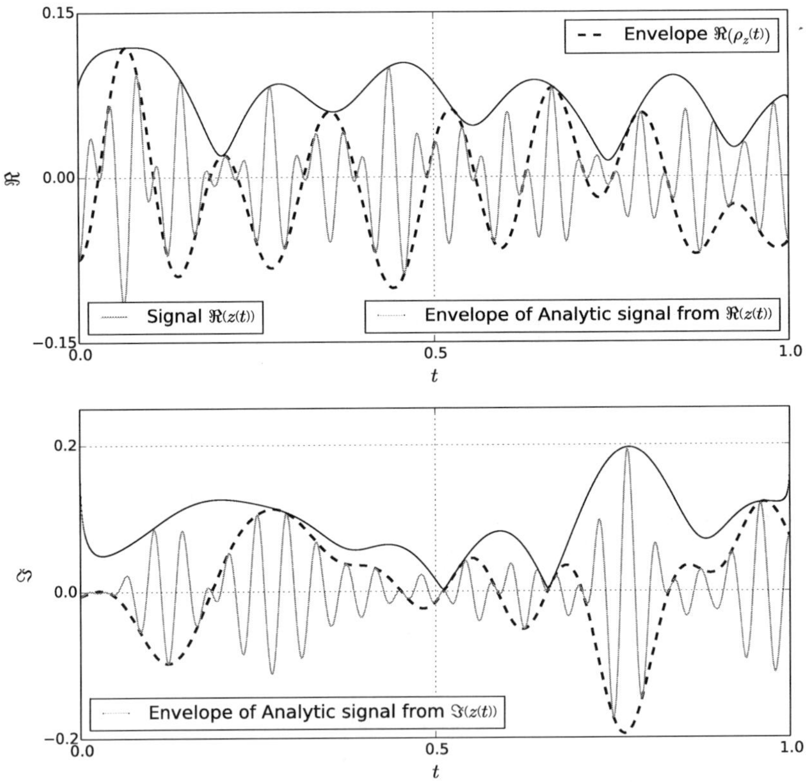

Figure 4.20. *Orthocomplex modulated complex signal. Real part (top) and imaginary part (bottom) of the complex signal $z(t)$ (gray solid lines) and its complex instantaneous amplitude $\rho_z(t)$ (dashed black lines) obtained from the CD polar of its quaternion extension $q_z(t)$. The envelopes computed using the "classical" analytic signal from $\Re(z(t))$ and $\Im(z(t))$ are displayed in black solid lines*

It must be noticed from the complex envelope displayed in Figures 4.19 and 4.20 that it is different from envelopes that could be obtained from the *classical* analytic signals computed separately on the real and imaginary parts of $z(t)$. This can be seen in Figure 4.20 where the envelopes obtained using the classical analytic signal computed on the real and imaginary parts of $z(t)$ separately are displayed. As both the envelopes computed using the classical analytic signal can only be positive valued (as they are moduli of complex-valued signals), there is no chance that they give an envelope that is a base-band version of the original complex signal $z(t)$. It must be emphasized that this is a major advantage of using the QFT to study complex-valued signals, as the complex envelope $\rho_z(t)$ has a direct interpretation, while any process computing the real and imaginary parts of $z(t)$ may not carry as much insight as the QFT approach does. Another approach to extend the concept of instantaneous attributes to complex signals has been proposed in [LIL 10]. It is a parametric approach called the *modulated elliptical signal* (MES), but it does not provide a complex-valued envelope for a given complex signal $z(t)$. A comparison of the MES approach with the instantaneous amplitude and frequency presented here is available in [LE 14].

4.3.6.2. *Example 2: orthocomplex modulated signal with linearly varying frequency*

In this example, we consider a complex signal with an instantaneous frequency linearly varying with time. The original $z(t)$ signal is displayed in Figure 4.21 in 3D and in Figure 4.22 as real and imaginary parts.

The frequency content of $z(t)$ is a single frequency that linearly changes with time. This can be guessed from Figures 4.21 and 4.22 where it can be seen that the frequency increases with time from $t = 0$ on, followed by a small decrease just before time $t = 1$.

The complex envelope $\rho_z(t)$ obtained from the polar CD form of the quaternion signal associated with $z(t)$ is displayed in 3D in Figure 4.21. In Figure 4.22, the real and imaginary parts of $z(t)$ are presented, together with the complex envelope. The envelopes computed using the classical analytic signal computed separately on the real and imaginary parts of $z(t)$ are superimposed. As we can see in Figure 4.22, the "classical" envelope cannot take negative values, which means that it cannot lead to the type of result obtained with $\rho_z(t)$. However, for positive values, the "classical" envelope and $\rho_z(t)$ take similar values.

Finally, to illustrate the instantaneous frequency concept, Figure 4.23 presents the estimated instantaneous frequency obtained by numerical differentiation of the phase $\Phi_z(t)$ given by the polar CD form of the quaternion signal $q_z(t)$. It must be noted that in theory, the phase $\Phi_z(t)$ could be complex-valued (see the definition in [1.60] and [1.61]). However, it can only take real values when the real and imaginary parts of the signal share the same frequency, which is the case here and is for most of the time in real-world applications. For example, it is well known that polarized signals can only be called so if both components of the electric field (classically denoted by $E_x(t)$ and $E_y(t)$) have frequencies in common [BRO 98]. As can be seen in Figure 4.23, the estimated frequency contains some errors. These errors occur at time instants where the signal $z(t)$ vanishes, leading to aberrant values in the phase estimation. However, the frequency content of the signal is recovered as can be seen from Figure 4.23 where the increasing frequency behavior together with the decrease at the end of the time window can be observed.

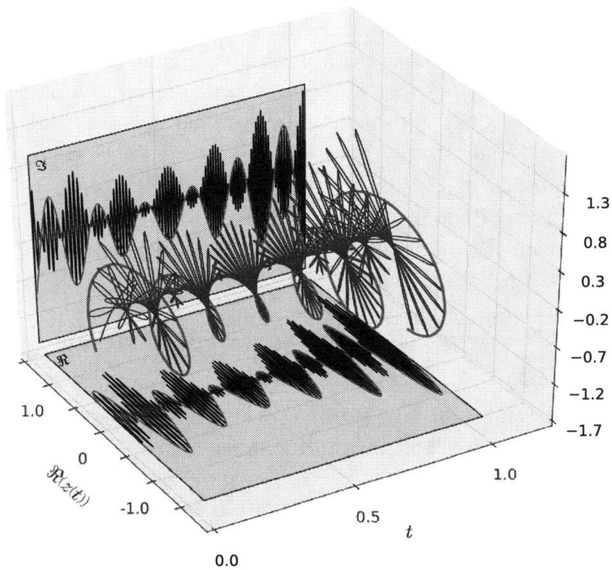

Figure 4.21. *Orthocomplex modulated complex signal with linearly varying frequency $z(t)$. 3D representation of the complex signal $z(t)$ (blue solid line) and its complex instantaneous amplitude $\rho_z(t)$ (red solid line) obtained from the polar CD form of its quaternion extension $q_z(t)$. The projections in $\Re$ and $\Im$ planes (black solid lines) correspond to the signals presented in Figure 4.22 (see color section within this book for a full color version of this figure)*

This demonstrates the ability of the polar CD form of the quaternion extension of a complex signal $z(t)$ to access its instantaneous frequency. This is a direct extension of what is known for real signals and the ability of the polar form of the analytic signal to provide the instantaneous phase and frequency.

Together, the instantaneous amplitude (envelope), the instantaneous phase and the instantaneous frequency obtained from the polar CD form of $q_z(t)$, the quaternion extension of the complex signal $z(t)$, are the faithful generalizations of the well-known instantaneous attributes obtained from the analytic signal of a real-valued signal. This shows the position of the QFT with respect to complex-valued signals: it is the counterpart of the (classical) FT for real signals.

This is due to the inherent symmetries of the QFT and its intrinsic four dimensions. The QFT should thus be considered on a systematic basis for the study of complex-valued signals as it has the required symmetries to extract any spectral information from bivariate signals.

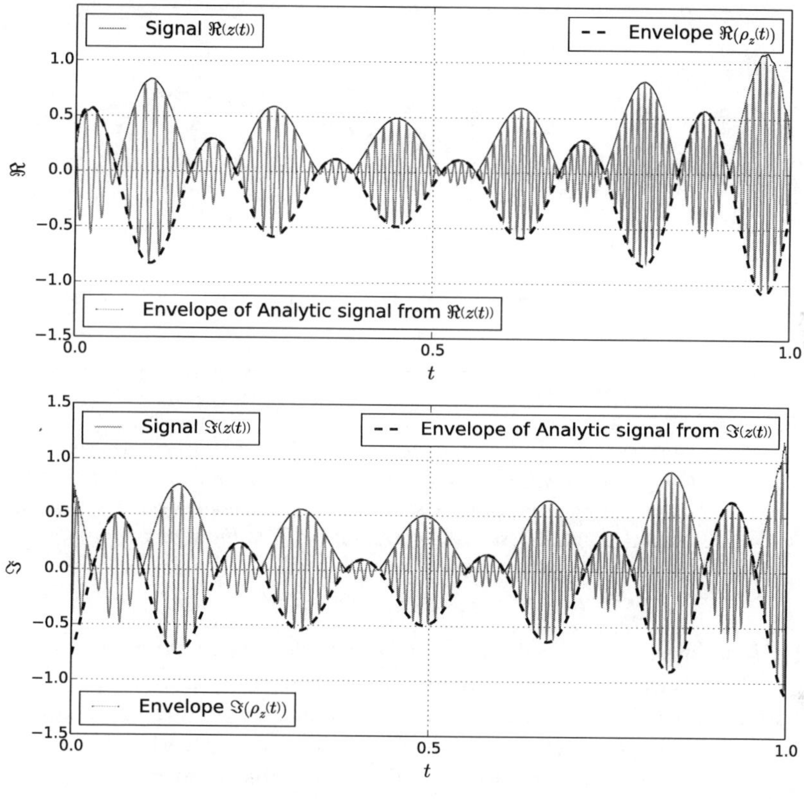

Figure 4.22. *Orthocomplex modulated complex signal. Real part (top) and imaginary part (bottom) of the complex signal $z(t)$ (gray solid lines) and its complex instantaneous amplitude $\rho_z(t)$ (dashed black lines) obtained from the CD polar form of its quaternion extension $q_z(t)$. The envelopes computed using the "classical" analytic signal from $\Re(z(t))$ and $\Im(z(t))$ are displayed as black solid lines*

4.3.7. *The quaternion Wigner-Ville distribution of a complex signal*

We end this section with some results that provide the basic foundations for a generalization of time-frequency representations inspired by the QFT[9].

9 A simple time-frequency analysis of non-stationary complex signals can also be carried out using a short-time QFT that can be easily computed using sliding time-windowed QFTs.

Time-frequency representations are well known in signal processing for the study of non-stationary signals [FLA 98]. Here, we show how the previously introduced concepts open the door to a new way to study non-stationary complex signals. Some parts of the presented material can be found in [ELL 12].

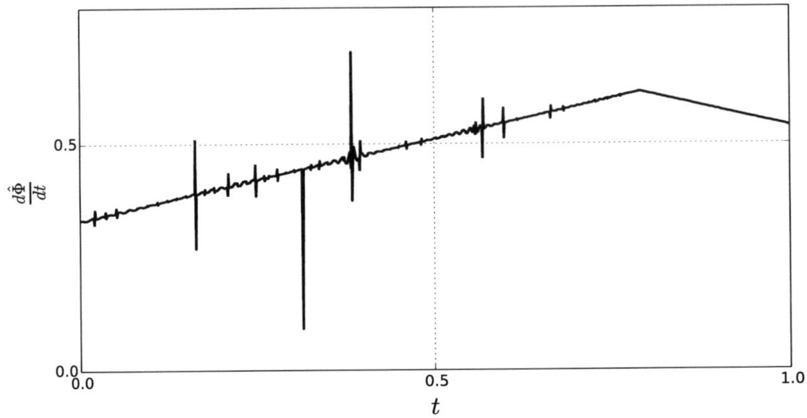

Figure 4.23. *Orthocomplex modulated complex signal with linearly varying frequency. Estimated instantaneous frequency obtained from the numerical derivative of the instantaneous phase computed as the phase of the polar CD form of $q_z(t)$, the quaternion signal associated with the complex signal $z(t)$. The estimation errors are due to the signal vanishing at certain times*

This section shows the link between the just defined *quaternion extension* and *instantaneous frequency* of a complex signal and its quaternionic Wigner-Ville distribution (QWVD). First, we introduce the definition of the quaternion Wigner distribution (QWD) of a complex signal and then the QWVD of its quaternion extension. Then, we introduce some of the remarkable properties of the QWVD. Finally, we show how the time marginal and mean frequency formulas [FLA 98] extend to the quaternionic case.

4.3.7.1. *Definitions*

Just as in the case of real signals, we will make the distinction between Wigner and the Wigner-Ville distribution. Recall that classically the Wigner distribution (WD) is built for real-valued signals, while the Wigner-Ville is built using the complex analytic signal associated with a real signal. The latter version was originally proposed by Ville [VIL 48]. Motivations for using Wigner-Ville distributions are given, for example, in [BOA 88].

First, we start with the Wigner distribution definitions. Given two complex-valued signals $z(t)$ and $w(t)$, their quaternionic (cross-) Wigner distribution with respect to the $\boldsymbol{j}$-axis, denoted by $W_{z,w}(t, \omega)$, is:

$$W_{z,w}(t,\omega) = \int_{-\infty}^{+\infty} z(t+\tau/2)w^{\star}(t-\tau/2)e^{-\boldsymbol{j}\omega\tau}\,\mathrm{d}\tau \qquad [4.42]$$

while the quaternionic (auto-) Wigner distribution of $z(t)$, denoted by $W_z(t, \omega)$, is given by:

$$W_z(t,\omega) = \int_{-\infty}^{+\infty} z(t+\tau/2)z^{\star}(t-\tau/2)e^{-\boldsymbol{j}\omega\tau}\,\mathrm{d}\tau \qquad [4.43]$$

Now, given two complex-valued signals $z(t)$ and $w(t)$ and their respective quaternion extensions $q_z(t)$ and $q_w(t)$, their quaternionic (cross-) Wigner-Ville distribution with respect to the $\boldsymbol{j}$-axis, denoted by $WV_{z,w}(t, \omega)$, is:

$$WV_{z,w}(t,\omega) = \int_{-\infty}^{+\infty} q_z(t+\tau/2)\overline{q_w}(t-\tau/2)e^{-\boldsymbol{j}\omega\tau}\,\mathrm{d}\tau \qquad [4.44]$$

while the quaternionic (auto-) Wigner-Ville distribution of $z(t)$, denoted by $WV_z(t, \omega)$, is given by:

$$WV_z(t,\omega) = \int_{-\infty}^{+\infty} q_z(t+\tau/2)\overline{q_z}(t-\tau/2)e^{-\boldsymbol{j}\omega\tau}\,\mathrm{d}\tau \qquad [4.45]$$

Note that in the Wigner-Ville definitions, the transform consists of a QFT with axis $\boldsymbol{j}$ of a quaternion-valued function with respect to the τ variable. Just as would happen with the complex FT of a complex signal, and because the quaternion extensions $q_z(t)$ and $q_w(t)$ have one-sided QFT spectra, the quaternionic auto- and cross-Wigner-Ville distributions are one-sided (i.e. equal to zero for negative frequencies).

In the sequel, we will consider Wigner and Wigner-Ville distributions, depending on the purpose we wish to illustrate. Mainly, we will stick to the Wigner case when dealing with standard properties, and then move to the Wigner-Ville case when considering the link with the instantaneous amplitude, phase and frequency of a complex signal.

4.3.7.2. *Basic properties*

Here, we list several properties of the QWD. Most of the proofs can be obtained by direct calculation, and can be found in [ELL 00a].

First, the QWD of the conjugate signal $z^\star(t)$ is simply:

$$W_{z^\star}(t,\omega) = W_z(t,\omega) \qquad [4.46]$$

and the (cross-) QWD fulfills the following symmetry relation when reversing the order of $z(t)$ and $w(t)$:

$$W_{w,z}(t,\omega) = \overline{W_{z,w}(t,\omega)}^{\boldsymbol{k}} \qquad [4.47]$$

The behavior of the QWD with respect to time shifting is as follows:

$$W_{\mathcal{T}_\chi z,\mathcal{T}_\chi w}(t,\omega) = W_{z,w}(t-\chi,\omega) \qquad [4.48]$$

where $\mathcal{T}_\chi z(t) = z(t-\chi)$ is a translation operator.

4.3.7.3. *Inversion*

It is possible to extend the well-known inversion formulas for the WD [FLA 98] to the quaternion version of the WD. Given two complex signals $z(t)$ and $w(t)$, the following equality holds:

$$z(t)w^\star(0) = \int_{-\infty}^{+\infty} W_{z,w}\left(\frac{t}{2},\omega\right) e^{\boldsymbol{j}\omega t}\,\mathrm{d}\omega \qquad [4.49]$$

provided that $w^\star(0) \neq 0$. Now, in the case of the (auto-) QWD of a complex signal $z(t)$, this becomes:

$$z(t)z^\star(0) = \int_{-\infty}^{+\infty} W_z\,(t,\omega)\,e^{\boldsymbol{j}\omega t}\,\mathrm{d}\omega \qquad [4.50]$$

This means that $z(t)$ can be recovered up to a phase factor $z^\star(0)$ by computing the inverse QFT of its QWD.

Now, we previously introduced the *orthocomplex modulation* which shares some behavior with the classical modulation when considering complex FTs. As was highlighted in section 4.3.1.2, the *orthocomplex modulation* is a central concept in the study of band-limited complex signals with right-sided quaternionic spectrum. It is thus interesting to know how this peculiar modulation behaves through the QWD, which is given, for two complex signals $z(t)$ and $w(t)$, by:

$$W_{\mathcal{O}_{\omega_0} z, \mathcal{O}_{\omega_0} w}(t, \omega) = W_{z,w}(t, \omega - \omega_0) \qquad [4.51]$$

where $[\mathcal{O}_{\omega_0} z](t) = z(t)e^{\jmath\omega_0 t}$ and $[\mathcal{O}_{\omega_0} w](t) = w(t)e^{\jmath\omega_0 t}$ defines the *orthocomplex modulation* operator.

In addition to the presented properties of the QWD, we now give the two characteristics that make the connection between the instantaneous phase and amplitude introduced in section 4.3.4 and some characteristics of the QWVD, namely the *time marginal* and the *mean frequency formula*.

4.3.8. *Time marginal*

We investigate here what information is available at each time instant t in the QWD of a complex signal $z(t)$ by integrating $W_z(t, \omega)$ over the frequency range. The result is summarized by the following expression:

$$\int_{-\infty}^{+\infty} W_z(t, \omega)\,\mathrm{d}\omega = z(t)z^\star(t) = |z(t)|^2 \qquad [4.52]$$

This is simply the squared instantaneous magnitude of the signal $z(t)$, i.e. its instantaneous energy. Doing the same with the QWVD would lead to the squared modulus of the quaternion extension of $z(t)$, which carries the same information in terms of energy. It would read as follows:

$$\int_{-\infty}^{+\infty} WV_z(t, \omega)\,\mathrm{d}\omega = q(t)\overline{q_z}(t) = |q_z(t)|^2 \qquad [4.53]$$

In the case of a real signal, it is well known that proceeding in the same way with the analytic signal associated with the original real signal leads to a similar result. Thus, the QWD (and QWVD) simply generalize this existing result to the case of an arbitrary complex signal.

4.3.9. *The mean frequency formula*

A more interesting result, which is known for the classical Wigner-Ville distribution, is that its mean value is the instantaneous frequency [FLA 98]. It states that the normalized mean value of the Wigner-Ville distribution is in fact the instantaneous frequency. We show here how this extends to the QWVD. The generalization leads to a more complicated expression for the mean value of the quaternion-valued QWVD; however, it is a generalization of the classical case and is given in [4.66].

As an illustration of calculations with a QFT, we give the lines to follow to obtain this property of the QWVD.

First, recall that if a complex signal $z(t)$ has a QFT denoted by $Z_{\boldsymbol{j}}(\omega)$, then the following property holds:

$$\mathcal{F}_{\boldsymbol{j}}\left\{\frac{\mathrm{d}}{\mathrm{d}t}z\right\}(\omega) = Z_{\boldsymbol{j}}(\omega)\boldsymbol{j}\omega \tag{4.54}$$

This extends to the QFT of the quaternion extension $q_z(t)$ directly by:

$$\mathcal{F}_{\boldsymbol{j}}\left\{\frac{\mathrm{d}}{\mathrm{d}t}q_z\right\}(\omega) = Z^+_{\boldsymbol{j}}(\omega)\boldsymbol{j}\omega \tag{4.55}$$

where we now have that $Z^+_{\boldsymbol{j}}(\omega) = 0$ for $\omega < 0$. We also have the classical *initial value* theorem that holds for the QFT of $q_z(t)$, i.e.:

$$\int_{-\infty}^{+\infty} Z^+_{\boldsymbol{j}}(\omega)\,\mathrm{d}\omega = q_z(t)|_{t=0} = q_z(0) \tag{4.56}$$

Now, from the definition of the QWVD, we can write that:

$$q_z\left(t+\frac{\tau}{2}\right)\overline{q_z}\left(t-\frac{\tau}{2}\right) \quad \xrightarrow{\mathcal{F}_j} \quad WV_z(t,\omega) \tag{4.57}$$

Now, remembering the above-mentioned property of the QFT with respect to derivation and that in the QWVD definition, integration is with respect to τ, we get:

$$\frac{\mathrm{d}}{\mathrm{d}\tau}\left(q_z(t+\frac{\tau}{2})\overline{q_z}(t-\frac{\tau}{2})\right) \quad \xrightarrow{\mathcal{F}_j} \quad WV_z(t,\omega)\boldsymbol{j}\omega \tag{4.58}$$

Now, expanding the derivative, denoting $q'_z(t) = \mathrm{d}q_z(t)/\mathrm{d}t$ and taking its value for $\tau = 0$, we get:

$$\int_{-\infty}^{+\infty} WV_z(t,\tau)\boldsymbol{j}\omega\,\mathrm{d}\omega = \frac{1}{2}\left[q'_z\left(t+\frac{\tau}{2}\right)\overline{q_z}\left(t-\frac{\tau}{2}\right) - q_z\left(t+\frac{\tau}{2}\right)\overline{q'_z}\left(t-\frac{\tau}{2}\right)\right]\Big|_{\tau=0} \tag{4.59}$$

Note that we left the integration over the range $]-\infty, +\infty[$ even though we know that the QWVD is right-sided. This does not affect the calculation anyway. Now, right-multiplying both sides of the equality by $\boldsymbol{j}$ leads to the following equality:

$$\int_{-\infty}^{+\infty} \omega WV_z(t,\omega)\,\mathrm{d}\omega = -\frac{1}{2}\left(q_z'(t)\overline{q_z}(t) - q_z(t)\overline{q_z'}(t)\right)\boldsymbol{j} \qquad [4.60]$$

Dividing both sides of the equality by $\int_{-\infty}^{+\infty} WV_z(t,\omega)\,\mathrm{d}\omega$ and remembering the result in [4.53], we get:

$$\begin{aligned}\frac{\int_{-\infty}^{+\infty} \omega WV_z(t,\omega)\,\mathrm{d}\omega}{\int_{-\infty}^{+\infty} WV_z(t,\omega)\,\mathrm{d}\omega} &= \frac{\frac{1}{2}\left(q_z'(t)\overline{q_z}(t) - q_z(t)\overline{q_z'}(t)\right)}{q_z(t)\overline{q_z}(t)}\boldsymbol{j} \\ &= -\frac{1}{2}\left(q_z'(t)q_z^{-1}(t) - \overline{q_z}^{-1}(t)\overline{q_z'}(t)\right)\boldsymbol{j} \\ &= -\mathbf{V}\left(q_z'(t)q_z^{-1}(t)\right)\boldsymbol{j}\end{aligned} \qquad [4.61]$$

It suffices now to remember that the quaternion extension $q_z(t)$ can be expressed with its polar CD form (as explained in [4.39]), so that $q_z(t) = \rho_z(t)e^{\phi_z(t)\boldsymbol{j}}$. This means that its time derivative is given by:

$$q_z'(t) = \rho_z'(t)e^{\phi_z(t)\boldsymbol{j}} + \rho_z(t)\phi_z'(t)\boldsymbol{j}e^{\phi_z(t)\boldsymbol{j}} = \left\{\rho_z'(t) + \rho_z(t)\phi_z'(t)\boldsymbol{j}\right\}e^{\phi_z(t)\boldsymbol{j}} \qquad [4.62]$$

Now, it follows that:

$$\begin{aligned}\frac{\int_{-\infty}^{+\infty} \omega\, WV_z(t,\omega)\,\mathrm{d}\omega}{\int_{-\infty}^{+\infty} WV_z(t,\omega)\,\mathrm{d}\omega} &= -\mathbf{V}\left(\frac{q_z'(t)}{q_z(t)}\right)\boldsymbol{j} \\ &= -\mathbf{V}\left(\frac{\rho_z'(t) + \rho_z(t)\phi_z'(t)\boldsymbol{j}}{\rho_z(t)}\right)\boldsymbol{j}\end{aligned} \qquad [4.63]$$

It is important to remember that $\rho_z(t)$ is complex-valued and that $\phi_z(t)$ can also be complex-valued (even if we already mentioned that it is most likely to be real-valued in real-world applications). Writing $\rho_z(t) = |\rho_z(t)|\exp\left(\varphi_{\rho_z(t)}\boldsymbol{i}\right)$, we first evaluate the quantity $q_z'(t)q_z^{-1}(t)$ as:

$$q_z'(t)q_z^{-1}(t) = \frac{|\rho_z(t)|'}{|\rho_z(t)|} + \boldsymbol{i}\varphi'_{\rho_z(t)} + \exp\left(\varphi_{\rho_z(t)}\boldsymbol{i}\right)\phi_z'(t)\boldsymbol{j} \qquad [4.64]$$

which is a full quaternion term as $|\rho_z(t)|' \,/\, |\rho_z(t)| \in \mathbb{R}$, and $\boldsymbol{i}\varphi'_{\rho_z(t)}$ is a pure $\Im_{\boldsymbol{i}}$ part and $\exp\left(\varphi_{\rho_z(t)}\boldsymbol{i}\right)\boldsymbol{j}$ is made of a $\Im_{\boldsymbol{j}}$ and a $\Im_{\boldsymbol{k}}$ part. Now, the final step consists of taking the vector part of this term and multiplying it on the right by $\boldsymbol{j}$, which leads to the following:

$$-\mathbf{V}\left(\frac{\rho_z'(t) + \rho_z(t)\phi_z'(t)\boldsymbol{j}}{\rho_z(t)}\right)\boldsymbol{j} = \exp\left(\varphi_{\rho_z(t)}\boldsymbol{i}\right)\phi_z'(t) - \boldsymbol{k}\varphi'_{\rho_z(t)} \quad [4.65]$$

which finally gives the following expression for the "mean frequency formula":

$$\frac{\int_{-\infty}^{+\infty} \omega WV_z(t,\omega)\,\mathrm{d}\omega}{\int_{-\infty}^{+\infty} WV_z(t,\omega)\,\mathrm{d}\omega} = \exp\left(\varphi_{\rho_z(t)}\boldsymbol{i}\right)\phi_z'(t) - \boldsymbol{k}\varphi'_{\rho_z(t)} \quad [4.66]$$

As said previously, this expression is more complicated than its "classical" version that involves the standard Wigner-Ville distribution of a real signal. Here, however, the instantaneous frequency of the complex signal $z(t)$ (defined as $\Omega_z(t) = \phi_z'(t)$ in section 4.3.5) can be recovered from the normalized mean of the QWVD. First, note that in [4.66], the right-hand side of the equation has two terms that do not mix together in the 4D quaternion space. The term $\boldsymbol{k}\varphi'_{\rho_z(t)}$ is a pure $\Im_k$ term that provides information on the complex envelope of $z(t)$. In fact, it is the instantaneous frequency of the complex envelope/amplitude $\rho_z(t)$ of $z(t)$. Next, the term $e^{\varphi_{\rho_z(t)}\boldsymbol{i}}\phi_z'(t)$ is made of the instantaneous frequency of $z(t)$, multiplied by a pure phase factor containing again information on the complex envelope. Thus, by simply taking the modulus of this term, we recover the instantaneous frequency $\phi'(t)$ of the complex signal $z(t)$. This is true in the cases where this instantaneous amplitude is real, which is, as already mentioned, the case in most real-world applications. The mean frequency formula thus provides information on the instantaneous frequency of the processed signal, just as in the classical case, and with some extra terms due to the additional information available in a complex signal (with respect to information available in a real-valued signal).

As a conclusion, we can see that the QWVD generalizes some well-known results from "classical" WVD, and may be a very powerful tool to analyze complex signals that arise in many applications. There is still some work to be done to completely discover the rest of the properties of the QWVD and of time-frequency representations based on quaternionic Fourier analysis.

Bibliography

[ALF 07] ALFSMANN D., GÖCKLER H., SANGWINE S.J. *et al.*, "Hypercomplex algebras in digital signal processing: benefits and drawbacks", *Proceedings of 15th European Signal Processing Conference (EUSIPCO 2007)*, Poznan, Poland, pp. 1322–1326, 3–7 September 2007.

[ALT 86] ALTMANN S.L., *Rotations, Quaternions and Double Groups*, Dover, 1986.

[BAE 02] BAEZ J.C., "The octonions", *Bulletin of the American Mathematical Society*, vol. 39, no. 2, pp. 145–205, 2002.

[BOA 88] BOASHASH B., "Note on the use of the Wigner distribution for time-frequency signal analysis", *IEEE Transactions on Acoustics, Speech and Signal Processing*, vol. 36, no. 9, pp. 1518–1521, 1988.

[BRA 74] BRANETS V.N. SHMYGLEVSKIY I.P., Application of Quaternions to Rigid Body Rotation Problems, NASA Technical Translation no. NASA TT F-15414, National Aeronautics and Space Administration, Washington, DC, July 1974, Translation of Primeneniye Kvaternionov v Zadachakh Orientatsii Tverdogo Tela, Nauka Press, Moscow, 1973. Available at http://archive.org/details/nasatechdoc19740020927.

[BRA 86] BRACEWELL R.N., *The Hartley Transform*, Oxford University Press, New York, 1986.

[BRA 00] BRACEWELL R.N., *The Fourier Transform and its Applications*, 3rd edition, McGraw–Hill, Boston, 2000.

[BRA 13] BRACKX F., HITZER E., SANGWINE S.J., "History of Quaternion and Clifford Fourier transforms and wavelets", in HITZER E., SANGWINE S.J. (eds), *Quaternion and Clifford Fourier Transforms and Wavelets*, Birkhäuser/Springer, pp. xi–xxvii, 2013.

[BRI 13] BRISTOW P.A., KORMANYOS C., HOLIN H., *et al.*, "Boost C++ Libraries, Math Toolkit", *Software Library*, 2006–2010, 2012, 2013. Available at http://www.boost.org/.

[BRO 98] BROSSEAU C., *Fundamentals of Polarized Light: A Statistical Optics Approach*, Wiley-Blackwell, 1998.

[BÜL 99] BÜLOW T., Hypercomplex spectral signal representations for the processing and analysis of images, PhD Thesis, University of Kiel, Germany, August 1999.

[BÜL 01] BÜLOW T., SOMMER G., "Hypercomplex signals a novel extension of the analytic signal to the multidimensional case", *IEEE Transactions on Signal Processing*, vol. 49, no. 11, pp. 2844–2852, November 2001.

[CHE 00] CHEN B., KAUFMAN A., "3D volume rotation using shear transformation", *Graphical Models*, vol. 62, pp. 308–322, 2000.

[CON 03] CONWAY J.H., SMITH D.A., *On Quaternions and Octonions*, A.K. Peters, Ltd., Natick, MA, 2003.

[COX 46] COXETER H.S.M., "Quaternions and reflections", *The American Mathematical Monthly*, vol. 53, no. 3, pp. 136–146, 1946.

[COX 74] COXETER H. S.M., *Regular Polytopes*, Dover Publications Inc., 1974.

[DEL 88] DELSUC M.A., "Spectral representation of 2D NMR spectra by hypercomplex numbers", *Journal of Magnetic Resonance*, vol. 77, no. 1, pp. 119–124, March 1988.

[DIC 14] DICKSON L., *Linear Algebras*, Cambridge University Press, London, 1914.

[DO 76] DO CARMO M.P., *Differential Geometry of Curves and Surfaces*, Pearson, 1976.

[ELL 00a] ELL T.A., SANGWINE S.J., "Decomposition of 2D hypercomplex Fourier transforms into pairs of complex Fourier transforms", in GABBOUJ M., KUOSMANEN P., (eds.), *Proceedings of EUSIPCO 2000, 10th European Signal Processing Conference*, Tampere, Finland, vol. 2, pp. 1061–1064, 5–8 September 2000.

[ELL 00b] ELL T.A., SANGWINE S.J., "Hypercomplex Wiener-Khintchine theorem with application to color image correlation", *IEEE International Conference on Image Processing (ICIP 2000)*, Vancouver, Canada, vol. 2, pp. 792–795, 11–14 September 2000.

[ELL 07a] ELL T.A., "Hypercomplex color affine filters", *IEEE International Conference on Image Processing (ICIP 2007)*, San Antonio, TX, vol. 5, pp. 249–252, 16–19 September 2007.

[ELL 07b] ELL T.A., On systems of linear quaternion functions, arXiv:math/0702084v1, February 2007.

[ELL 07c] ELL T.A., SANGWINE S.J., "Hypercomplex Fourier transforms of color images", *IEEE Transactions on Signal Processing*, vol. 16, no. 1, pp. 22–35, January 2007.

[ELL 07d] ELL T.A., SANGWINE S.J., "Quaternion involutions and anti-involutions", *Computers and Mathematics with Applications*, vol. 53, no. 1, pp. 137–143, 2007.

[ELL 12] ELL T.A., LE BIHAN N., "Quaternion Wigner-Ville distribution", *Applied Geometric Algebras in Computer Science and Engineering (AGACSE)*, La Rochelle, France, 2012.

[ERN 87] ERNST R.R., BODENHAUSEN G., WOKAUN A., *Principles of Nuclear Magnetic Resonance in One and Two Dimensions*, Oxford University Press, Oxford, 1987.

[FLA 98] FLANDRIN P., *Time-Frequency/Time-Scale Analysis*, Academic Press, 1998.

[FRI 05] FRIGO M., JOHNSON S.G., "The design and implementation of FFTW3", *Proceedings of the IEEE*, vol. 93, no. 2, pp. 216–231, 2005. Special issue on "Program generation, optimization, and platform adaptation".

[GAB 46] GABOR D., "Theory of communication", *Journal of the IEE*, vol. 93, no. 26, pp. 429–457, 1946.

[GIL 08] GILMORE R., *Lie Groups, Physics and Geometry*, Cambridge University Press, 2008.

[GOL 96] GOLUB G.H., VAN LOAN C.F., *Matrix Computations*, The Johns Hopkins University Press, 3rd edition, Baltimore and London, 1996.

[HAN 06] HANSON A.J., *Visualizing Quaternions*, Morgan Kaufmann, 2006.

[HIT 07] HITZER E., "Quaternion Fourier transform on quaternion fields and generalizations", *Advances in Applied Clifford Algebras*, vol. 17, no. 3, pp. 497–517, 2007.

[HIT 13] HITZER E., SANGWINE S.J., Eds., *Quaternion and Clifford Fourier Transforms and Wavelets*, Birkhäuser/Springer, Basel, Switzerland, 2013.

[ICK 70] ICKES B.P., "A new method for performing digital control system attitude computations using quaternions", *AIAA Journal*, vol. 8, no. 1, pp. 13–17, January 1970.

[JOL 05] JOLY C.J., *A Manual of Quaternions*, Macmillan, 1905.

[KAN 89] KANTOR I., SOLODOVNIKOV A., *Hypercomplex Numbers: An Elementary Introduction to Algebras*, Springer-Verlag, 1989.

[KUG 75] KUGLIN C.D., HINES D.C., "The phase correlation image alignment method", *Proceedings of IEEE Conference on Cybernetics and Society*, San Francisco, CA, IEEE, pp. 163–165, 23–25 September 1975.

[KUI 02] KUIPERS J., *Quaternions and Rotation Sequences: A Primer with Applications to Orbits, Aerospace and Virtual Reality*, Princeton University Press, 2002.

[LE 14] LE BIHAN N., SANGWINE S.J., ELL T.A., "Instantaneous frequency and amplitude of orthocomplex modulated signals based on quaternion Fourier transform", *Signal Processing*, vol. 94, pp. 308–318, January 2014.

[LIL 10] LILLY J.M., OLHEDE S.C., "Bivariate instantaneous frequency and bandwidth", *IEEE Transactions on Signal Processing*, vol. 58, no. 2, pp. 591–603, 2010.

[LIL 11] LILLY J.M., "Modulated oscillations in three dimensions", *IEEE Transactions on Signal Processing*, vol. 59, no. 12, pp. 5930–5943, 2011.

[LOU 01] LOUNESTO P., *Clifford Algebras and Spinors*, Cambridge University Press, Cambridge, 2001.

[MAC 06] MACFARLANE A., *Vector Analysis and Quaternions*, John Wiley & Sons, New York, 1906.

[MCI 98] MCILHAGGA W., "Colour vision", in SANGWINE S.J., HORNE R.E.N. (eds), *The Colour Image Processing Handbook*, Chapter 2, Chapman and Hall, pp. 7–25, 1998.

[MES 83] MESERVE B.E., *Fundamental Concepts of Geometry*, Corrected reprint of 1959 edition, Dover, New York, 1983.

[MOX 03] MOXEY C.E., SANGWINE S.J., ELL T.A., "Hypercomplex correlation techniques for vector images", *IEEE Transactions on Signal Processing*, vol. 51, no. 7, pp. 1941–1953, July 2003.

[PAL 98] PALUS H., "Representations of colour images in different colour spaces", in SANGWINE S.J., HORNE R.E.N. (eds), *The Colour Image Processing Handbook*, Chapter 4, Chapman and Hall, pp. 67–90, 1998.

[PIC 97] PICINBONO B., "On instantaneous amplitude and phase of signals", *IEEE Transactions on Signal Processing*, vol. 45, no. 3, pp. 552–560, 1997.

[POR 95] PORTEOUS I.R., *Clifford Algebras and the Classical Groups*, Cambridge University Press, 1995.

[PRI 91] PRICE G.B., *An Introduction to Multicomplex Spaces and Functions*, Marcel Dekker Inc., New York, 1991.

[ROB 68] ROBINSON P.D., *Fourier and Laplace Transforms*, Routledge and Kegan Paul, London, 1968.

[ROC 04] ROCHON D., TREMBLAY S., "Bicomplex quantum mechanics: I. the generalized Schrödinger equation", *Advances in Applied Clifford Algebras*, vol. 14, no. 2, pp. 231–248, 2004.

[ROW 14] ROWLAND T., WEISSTEIN E.W., "Matrix exponential", *MathWorld*. Available at Wolfram Web Resource, http://mathworld.wolfram.com/ MatrixExponential.html, January 2014.

[SAN 96] SANGWINE S.J., "Fourier transforms of colour images using quaternion, or hypercomplex, numbers", *Electronics Letters*, vol. 32, no. 21, pp. 1979–1980, 10 October 1996.

[SAN 98a] SANGWINE S.J., HORNE R.E.N., (eds.), *The Colour Image Processing Handbook*, Chapman and Hall, London, 1998.

[SAN 98b] SANGWINE S.J., THORNTON A.L., "Frequency domain methods", in SANGWINE S.J., HORNE R.E.N. (eds), *The Colour Image Processing Handbook*, Chapter 12, Chapman and Hall, pp. 228–241, 1998.

[SAN 98c] SANGWINE S.J., "Colour image edge detector based on quaternion convolution", *Electronics Letters*, vol. 34, no. 10, pp. 969–971, 14 May 1998.

[SAN 00] SANGWINE S.J., ELL T.A., "Colour image filters based on hypercomplex convolution", *IEE Proceedings – Vision, Image and Signal Processing*, vol. 147, no. 2, pp. 89–93, April 2000.

[SAN 01] SANGWINE S.J., ELL T.A., MOXEY C.E., "Vector phase correlation", *Electronics Letters*, vol. 37, no. 25, pp. 1513–5, 6 December 2001.

[SAN 04] SANGWINE S.J., ELL T.A., "Gray-centered RGB color space", *Second European Conference on Color in Graphics, Imaging and Vision (CGIV 2004)*, Technology Center AGIT, Aachen, Germany, pp. 183–186, 5–8 April 2004.

[SAN 10] SANGWINE S.J., LE BIHAN N., "Quaternion polar representation with a complex modulus and complex argument inspired by the Cayley-Dickson form", *Advances in Applied Clifford Algebras*, vol. 20, no. 1, pp. 111–120, March 2010.

[SAN 12] SANGWINE S.J., ELL T.A., "Complex and hypercomplex discrete Fourier transforms based on matrix exponential form of Euler's formula", *Applied Mathematics and Computation*, vol. 219, no. 2, pp. 644–655, October 2012.

[SAN 13a] SANGWINE S.J., "Perspectives on color image processing by linear vector methods using projective geometric transformations", in HAWKES P.W. (ed.), *Advances in Imaging and Electron Physics*, Academic Press: Elsevier Inc., vol. 175, pp. 283–307, 2013.

[SAN 13b] SANGWINE S.J., LE BIHAN N., "Quaternion Toolbox for Matlab®, Version 2 with support for Octonions", *Software Library*, 2013. Available at http://qtfm.sourceforge.net/.

[SHO 85] SHOEMAKE K., "Animating rotation with quaternion curves", *SIGGRAPH*, vol. 19–3, pp. 245–254, 1985.

[SNE 61] SNEDDON I.N., *Fourier Series*, Routledge and Kegan Paul, London, 1961.

[SOM 01] SOMMER G., (ed.), *Geometric Computing with Clifford Algebras: Theoretical Foundations and Applications in Computer Vision and Robotics*, Springer-Verlag, London, UK, 2001.

[STI 10] STILLWELL J., *Mathematics and Its History*, Undergraduate Texts in Mathematics, 3rd edition, Springer Science+Business Media, LLC, 2010.

[SYN 72] SYNGE J.L., *Quaternions, Lorentz Transformations, and the Conway-Dirac-Eddington Matrices*, Communications of the Dublin Institute for Advanced Studies, Series A no. 21, Dublin Institute for Advanced Studies, Dublin, 1972.

[THO 98] THORNTON A.L., Colour object recognition using a complex colour representation and the frequency domain, PhD Thesis, Department of Engineering, The University of Reading, 1998.

[TRU 53] TRUMPLER R.J., WEAVER H.F., *Statistical Astronomy*, University of California Press, Berkeley and Los Angeles, CA, 1953.

[VIL 48] VILLE J., "Théorie et Applications de la Notion de Signal Analytique", *Cables et Transmission*, vol. 2A, pp. 61–74, 1948.

[WAR 97] WARD J.P., *Quaternions and Cayley Numbers: Algebra and Applications*, Kluwer, Dordrecht, vol. 403, 1997.

[WEI 14a] WEISSTEIN E.W., "Cross-correlation", *MathWorld.* Available at Wolfram Web Resource, http://mathworld.wolfram.com/Cross-Correlation.html, January 2014.

[WEI 14b] WEISSTEIN E.W., "Cross-correlation theorem", *MathWorld.* Available at Wolfram Web Resource, http://mathworld.wolfram.com/Cross-CorrelationTheorem.html, January 2014.

[YEH 08] YEH M.-H., "Relationships among various 2-D quaternion Fourier transforms", *Signal Processing Letters, IEEE*, vol. 15, pp. 669–672, 2008.

[ZHA 97] ZHANG F.Z., "Quaternions and matrices of quaternions", *Linear Algebra and its Applications*, vol. 251, pp. 21–57, January 1997.

Index

A

absolutely integrable, 40
algebra
  Clifford, 65
  division, 1, 3, 5
amplitude
  instantaneous, 101
analytic signal, 96, 100
anti-involution, 4, 23
arc, spherical, 27
associativity, 3
axis, 11

B

Baker-Campbell-Hausdorff formula, 9
basis, 2
  change of, 61
bicomplex product
  commutativity, 16
binormal vector, 103
bivariate signal analysis, 91
Boost, 57

C

Cayley-Dickson form, 12
  polar, 14
change of basis, 61
change of coordinates, 2D, 52
$\mathbb{C}_j$-pair, 15
color image, 70
  decomposition, 75
color space
  RGB, 70
commutativity
  non-, 3
complex exponential, 82
complex signal
  instantaneous amplitude, 91
  instantaneous phase, 91
  properties of 1D QFT, 91
complex subfield, 86
conjugate, 4
  time/frequency reversal, 45
conjugation
  quaternion, 23
convolution, 67
  classical, 2D, 67
  complex, 36, 96
  mask, 68
  quaternion, 70
  theorem, 47, 68
coordinates
  change of, 52
  homogeneous, 28
correlation
  classical, 81
  color and grayscale
    image, 91
  complex, 36
  frequency domain, 86
  peak interpretation, 88

phase, 80–82, 88
sub-sample precision, 84
quaternion, 79, 86
theorem, 47
cross product, 3
cross-covariance, 81
cross-power spectrum, 80

D

de Moivre, 18
decomposition
symplectic, 45
decompositions, 42
2D, 54
derivatives, 42, 53
dilation, 3D, 24
dirac delta function, 42
dirichlet conditions, 39
division algebra 1, 3, 5

E

edge detector, 68
envelope, 103
complex, 105, 107
Euler angles, 12
formula, 8, 11, 64
exponential
complex, 82
matrix, 63
quaternion, 7, 9

F

FFTW library, 61
Fourier
convolution theorem, 68
transform
coefficients, 65
complex, 35, 37
computation by decomposition, 60
direct DFT coding, 58
direct FFT coding, 60
discrete-time, 65
dual-axis form, 48
factored form, 48
fast, 60
FFT, 60
matrix exponential, 63
pairs, 1D, 40
pairs, 2D, 56
pairs, complex, 37
quaternion, 1D, 38
quaternion, 2D, 48
tiling, 82
verification, 62
transforms
Clifford
coding, 57
computation of, 57
Frenet-Serret formulas, 103
frequency
instantaneous, 91
mean
Wigner-Ville distribution, 113
negative, 66, 80, 97
Nyquist, 66, 75
frequency domain, 81

G

Gabor, D., 96
geometry, 21
Euclidean, 21
spherical, 26
great circle, 27
group
general linear, 32, 33
group velocity, 91

H

Hamilton, Sir W. R., 1
Hilbert transform, 96, 102
homogeneous coordinates, 28

I

image
color, 70
grayscale, 67
registration, 91
impulse, 42, 53
instantaneous
amplitude, 91, 101
frequency, 91
of a complex signal, 102

phase, 91, 101
integrable absolutely, 40
involution, 4, 5, 15
isometry, 33

K, L, M

Kirsch, 68
linearity, 40, 52
logarithm
quaternion, 7
marginal processing, 84
MATLAB®, 57, 64
matrix
exponential, 63
representations, 17
mean frequency
Wigner-Ville distribution, 113
modulated elliptical signal, 107
modulation
orthocomplex,
95, 112
modulo arithmetic, 82
modulus, 4

N, O

norm 1
product of, 3
quaternion, 3
Nyquist frequency, 66, 75
one-sided spectrum, 97
orthocomplex modulation, 95, 112
orthogonal split, 2D, 45
osculating plane, 103

P

phase
Fourier coefficients, 80
instantaneous, 101
phase correlation, 80, 82
classical, 81
phase velocity, 91
plane, osculating, 103
point, weighted, 28
polar form, 101
Cayley-Dickson, 102
Prewitt, 68
product
scalar, 3
vector, 3
projective space, 29
(3D), 28

Q

QTFM, 57
quadrature, 105
signal, 96
quaternion
axis, 11
basis, 2, 6
Cartesian form, 1
Cayley-Dickson form, 12
polar, 14
$\mathbb{C}_{\boldsymbol{j}}$-pair, 15
conjugate, 4
convolution, 70
de Moivre formula, 18
exponential, 7, 9
Fourier
transform, *see* Fourier
functions
linear, 31
$\boldsymbol{ijk}$, 1
inverse, 5
involution, 5, 15
properties, 5
logarithm, 7, 10
modulus, 4
norm, 3, 4
perplex part, 13
polar form, 8, 11
Euler angles, 12
power, 18
product, 3
bicomplex, 15
matrix-vector form, 32
pure, 2
ratio of, 5
representations, 11
matrix, 17
scalar part, 1
simplex part, 13
subfields, 18
sum, 2

symplectic form, 13
systems of equations, 31
vector part, 1

R

reflection
3D, 22
4D, 25
registration
image, 91
reversal
time, 80
time/frequency, 45, 46
implementation, 65
rotation, 71
3D, 22
4D, 25
composition, 23

S

scalar part, 1
using conjugate, 4
scalar product, 3
scaling, 40, 52
shear, 24
shift
frequency, 95
shifting, 40, 53
in correlation, 81
signal
analytic, 100
complex
instantaneous frequency, 102
non-stationary, 109
quaternion extension, 101
envelope, 103
quadrature, 96, 98
quaternion
associated with complex, 98
signal analysis
bivariate, 91
Sobel, 68
spectrum
cross-power, 80
one-sided, 97
quaternion
layout, 75
visualization, 74
quaternion image, 73
spherical triangle, 28
split
even-odd, 42
orthogonal 2D plane, 45
subfield
complex, 86
superposition, 33
swap rule, 13
symmetry, 42, 75, 80, 94, 100
complex signal, 93
QFT, 109
QFT of complex signal
properties of, 98
symplectic decomposition, 45
symplectic form, 13
Synge, J.L., 81

T

tangent vector, 103
tiling, 82
time domain, 80
time reversal, 80
time-frequency analysis, 91, 101
time/frequency reversal, 65
transform
Hilbert, 96, 102
transformation
affine, 33
dilation, 24
Euclidean, 33
linear, 33
projective, 33
reflection
3D, 22
4D, 25
rotation, 71
3D, 22
4D, 25
shear, 24
similarity, 33
translation vector, 28
transmutation, 32
triangle, spherical, 28

V

vector
    translation, 28
vector part
    using conjugate, 4
versor, 6, 10
Ville, J., 96
visual system
    human, 70
visualization, quaternion spectrum, 74

W, Z

Wigner-Ville distribution
    mean frequency, 113
    quaternion, 109, 112
zero padding, 82